TUMI

TUMI

The TUMI Collection

Foreword by Matt Hranek
Introduction by Victor Sanz

Rizzoli
New York
New York · Paris · London · Milan

Profiles

Foreword by Matt Hranek

Editor

I have traveled quite a bit. As a magazine editor I have had the privilege to navigate many airports and hotels throughout my career and I have noticed a lot of TUMI luggage. It is easy to spot the clean-designed lines of the bags and the bright red branding of the TUMI logo. I run into many travelers at airline check in lines with well-worn and patinaed TUMI bags that have a great history and the potential for stories to tell about life on the road. I also see quite a few backpacks being stuffed into over head bins and under seats, many are TUMI. I think it's fair to say that there is a lot of TUMI being used by a dedicated, brand-loyal group of travelers from businessmen to seasoned professionals. That was the inspiration for this book. We knew that TUMI had a prolific following of travelers, professionals, and celebrities, and we vetted a great cross section of these TUMI lovers to tell their stories.

We thought this was going to be a travel journal of sorts. Owners would be interviewed and tell their tales of great trips and share how TUMI played a part on that journey. What I learned after dozens of interviews was something a bit different.

My wife and I spent the drive to our country house some three hours from New York City listening to many of the TUMI interviews that I had recorded. After hours of listening to celebrities, photographers, musicians, stylists, and even a pro surfer, I realized there was a common thread.

Sure all these professionals liked the overall simple and elegant aesthetics of TUMI. Yet moreover, they saw TUMI as an essential part of their travel tools and an intricate part of their professional life and craft. After listening to so many inspiring, intimate TUMI stories, my wife turned to me and said, “I have to own a TUMI.” That is the power of the brand. Some people spent more time with their TUMI than their families. Some gave their bag a special name and all cherished them as an heirloom of sorts. Choosing not to ever part with their TUMI but rather keeping them to hand down to the next generation. I've experienced this kind of attachment with watches and even cars but luggage was a new category for me.

There are a few things I've learned with this project. One is, TUMI is definitely part of my travel arsenal and now by simply rolling my bag in the airport to get to my gate, I am a part of a bigger community. A community that shares a common love for an object that has become bigger than the sum of its parts.

Introduction by Victor Sanz

TUMI Creative Director

The TUMI Collection is dedicated to our incredible global community.

The movers and makers who travel the globe in pursuit of their passions are the inspiration behind everything we do. We are dedicated to upgrading, uncomplicating, and beautifying all aspects of life on the move. For us, the pursuit of excellence is never-ending.

From the inception of the TUMI brand, the needs of our global citizens have always played a pivotal role in how we create and communicate. Getting a chance to hear their amazing stories over the years, and figuring out ways to make their adventures even more memorable, drives us every day.

So, when it came time to put pen to paper, it was clear that our customers should be the ones to help us tell the story. A collection of their stories is, in fact, the best way to tell the story of our daily pursuit, and the reason why I have loved being a part of the TUMI brand for the past 17 years.

From being a part of the Kravitzs' first father-daughter trip to the place where Lenny's mother grew up, to collaborating with Vice and Russell Westbrook to bring their visions to life, TUMI is a reflection of not only my personal and professional values and passions, but also those of our global citizens. Each person in this book has a remarkable story. These stories, and many others, have helped shaped this brand into what it is today and continue to be a vital part of TUMI's evolution.

We hope you enjoy reading it as much as we enjoyed the process of creating it.

The most important thing for me, when I'm traveling, is to feel comfortable.

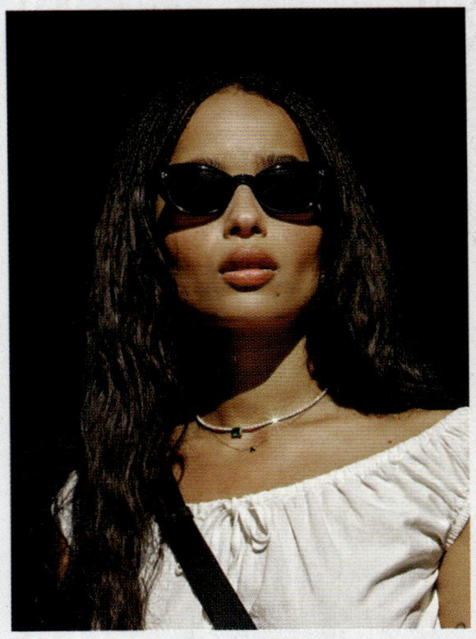

Zoë Kravitz

Actress and Musician

The more I travel, the more I appreciate TUMI. I used to throw everything into whatever bag I could find. But, as someone who basically lives at the airport, TUMI has made my life so much easier. The carry-ons, with all those compartments, are so practical. I think they really help me with my entire packing process. And it's also really great to have luggage that's so nice to look at. For a long time, I didn't really think of luggage as an accessory, like a purse or a pair of shoes. I only thought of how it functioned. But now that I travel so much, it feels great to grab something that's so simple, yet elegant. I always pack things that make me feel at home, like a candle or incense or a soft blanket. I think there's something really vulnerable about travel. You have to bring your comfort. You have to bring the things that remind you of home.

Collection
Voyageur

Style
Ruma Crossbody

Collection Launched
2019

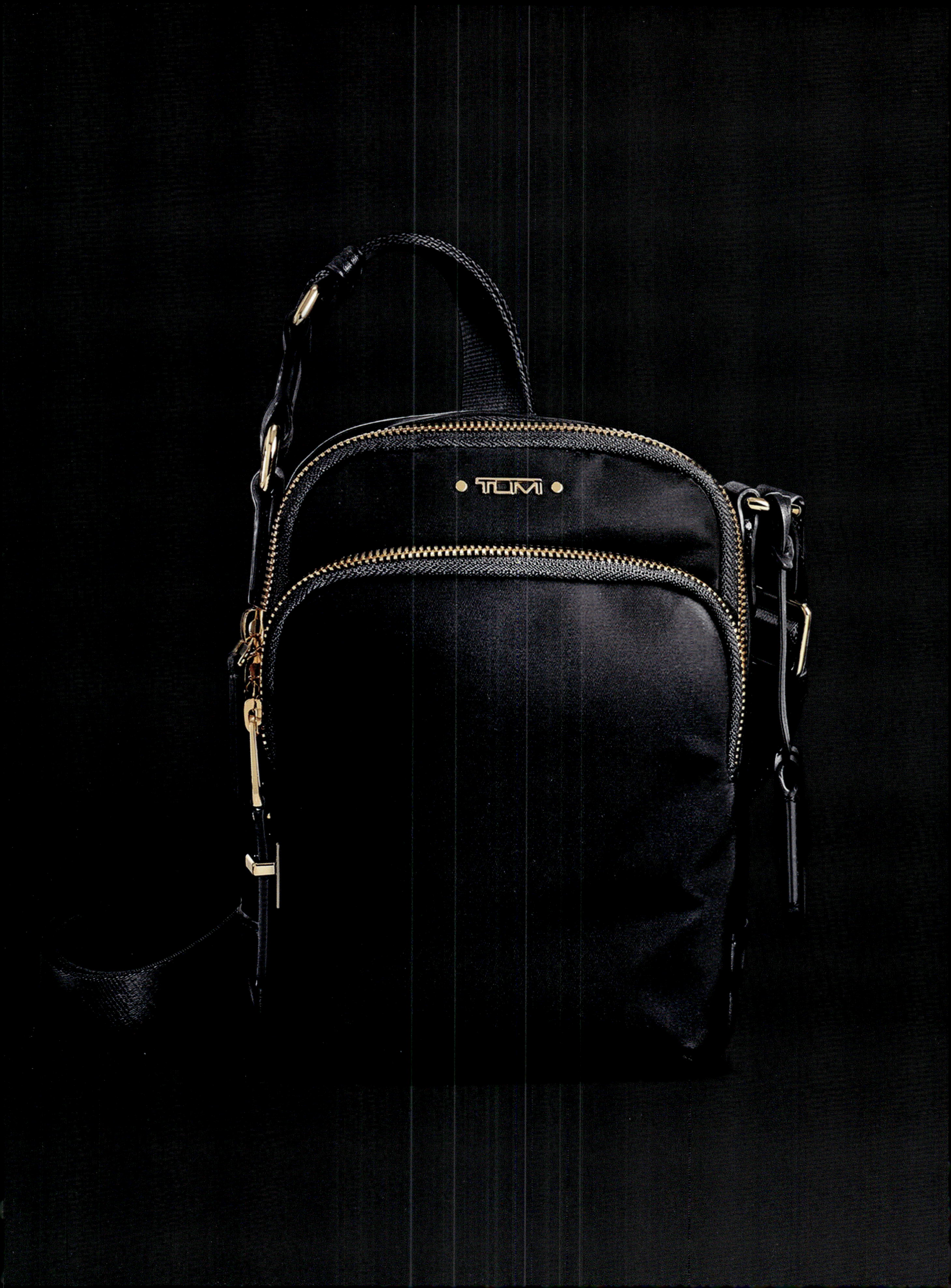
TUMI

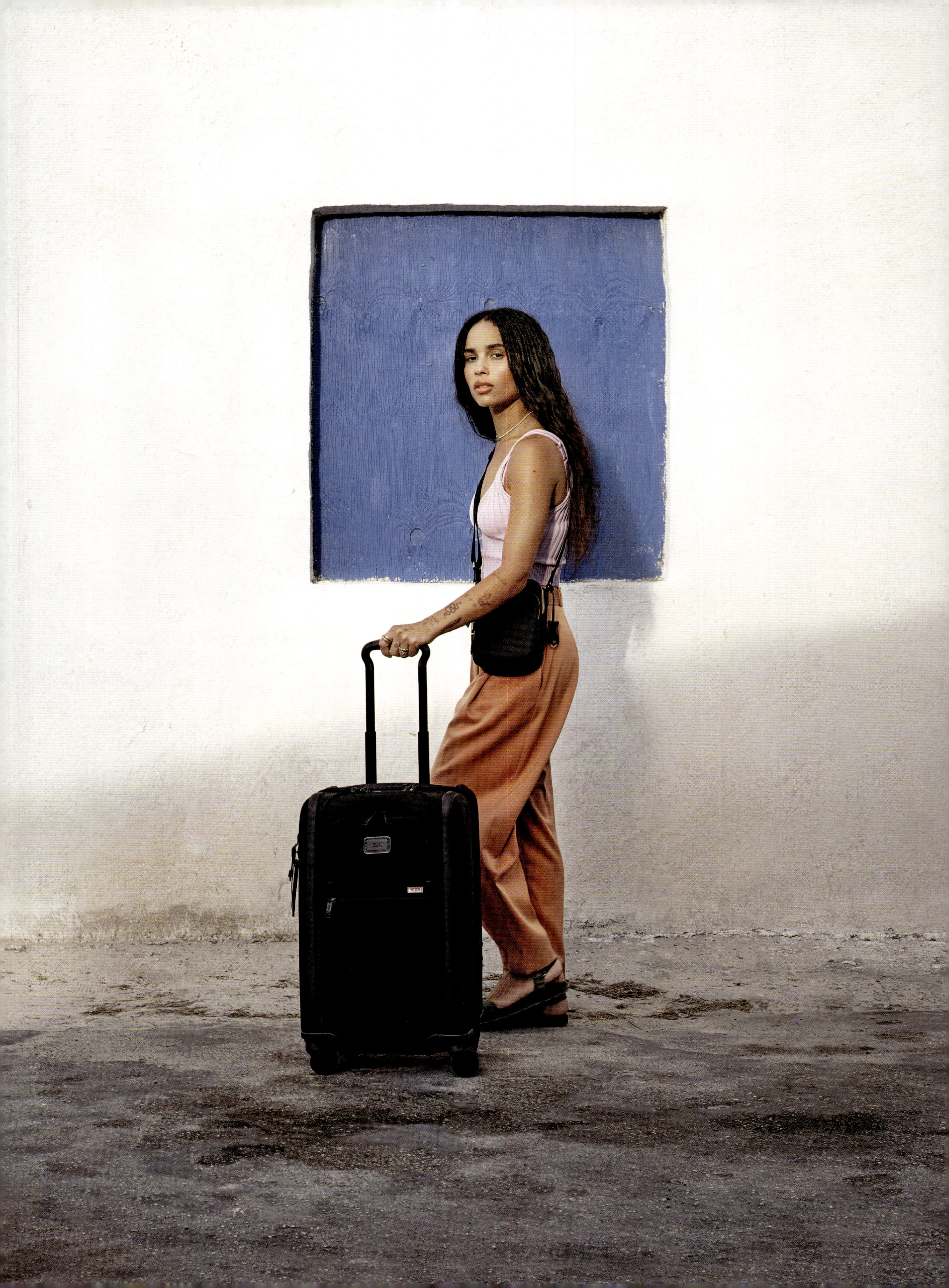

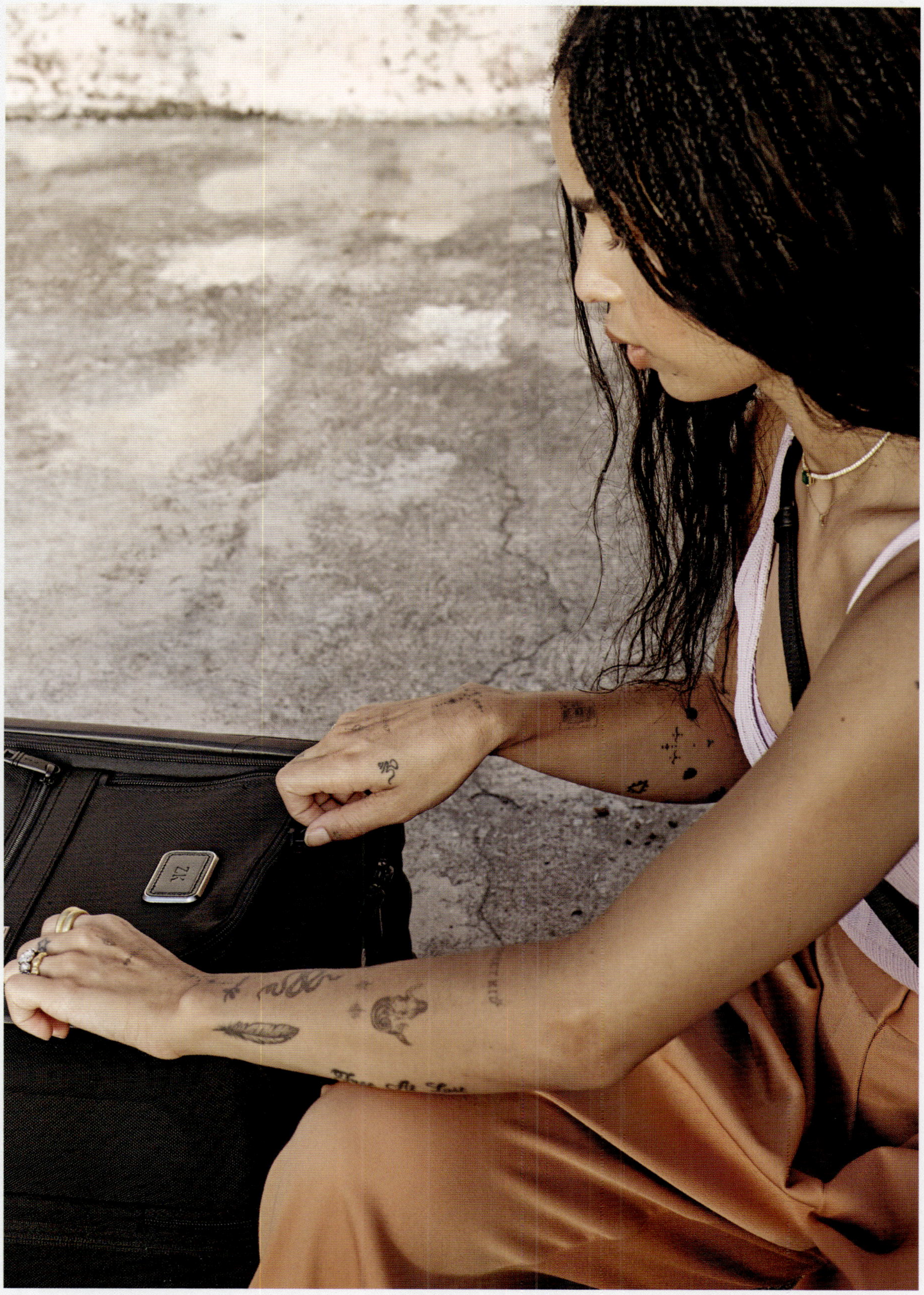

Collection

Vapor Lite

Style

Extended Trip
Packing Case

Collection
Launched

2015

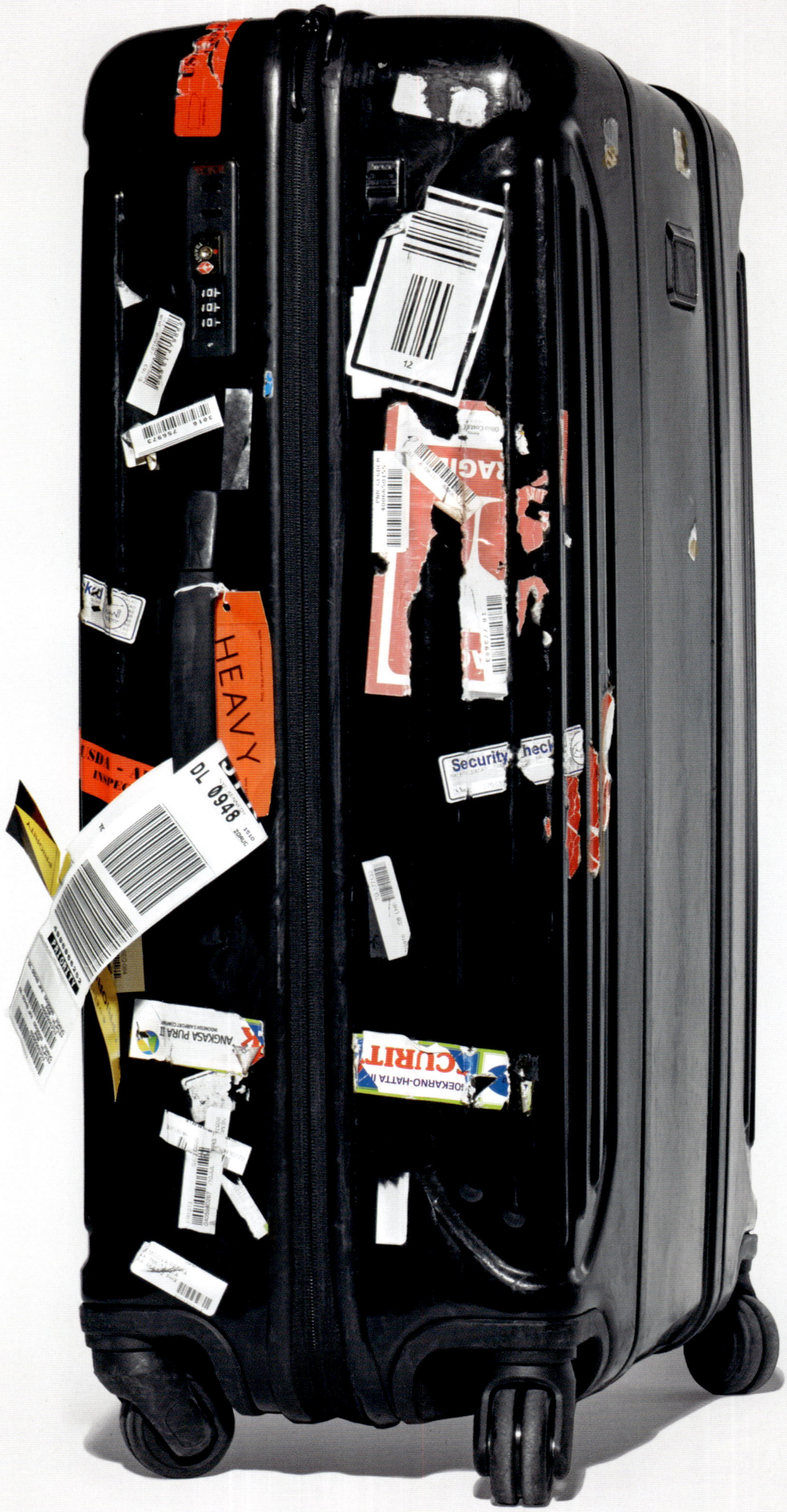

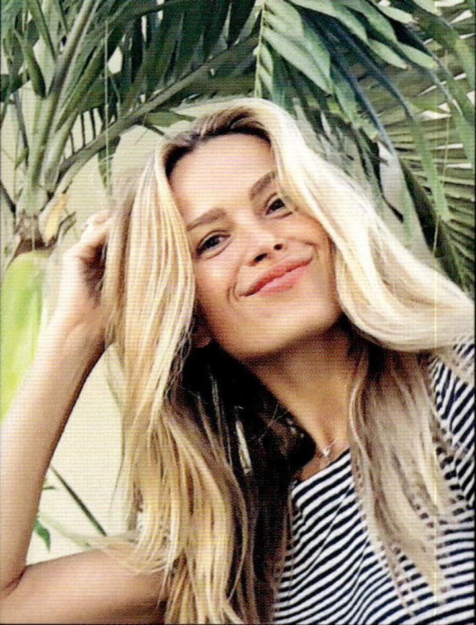

Petra Němcová

Model, Activist, and Co-Founder of All Hands and Hearts

I spend most of my time on the road.

It's not uncommon for me to be on a plane upwards of ten times per week, so I really do see my luggage as my mobile home. In my backpack, for example, that's where I keep my cosmetics, supplements, and vegan snacks. In another part of the backpack, I'll pack my books, documents, and electronics. Everything has its own perfect place. If I'm in a pitch dark room, I can reach into my backpack and know exactly where to find everything I need.

Each bag represents a part of home that I carry with me, and because of that I'm quite attached to my luggage. My roller has been through the trenches—the pushing, pulling, and more than a few times—when it's been lost en route—it's literally chased me around the world. This suitcase, which is now totally covered in barcodes and FRAGILE stickers, has been through a lot. This suitcase has truly lived.

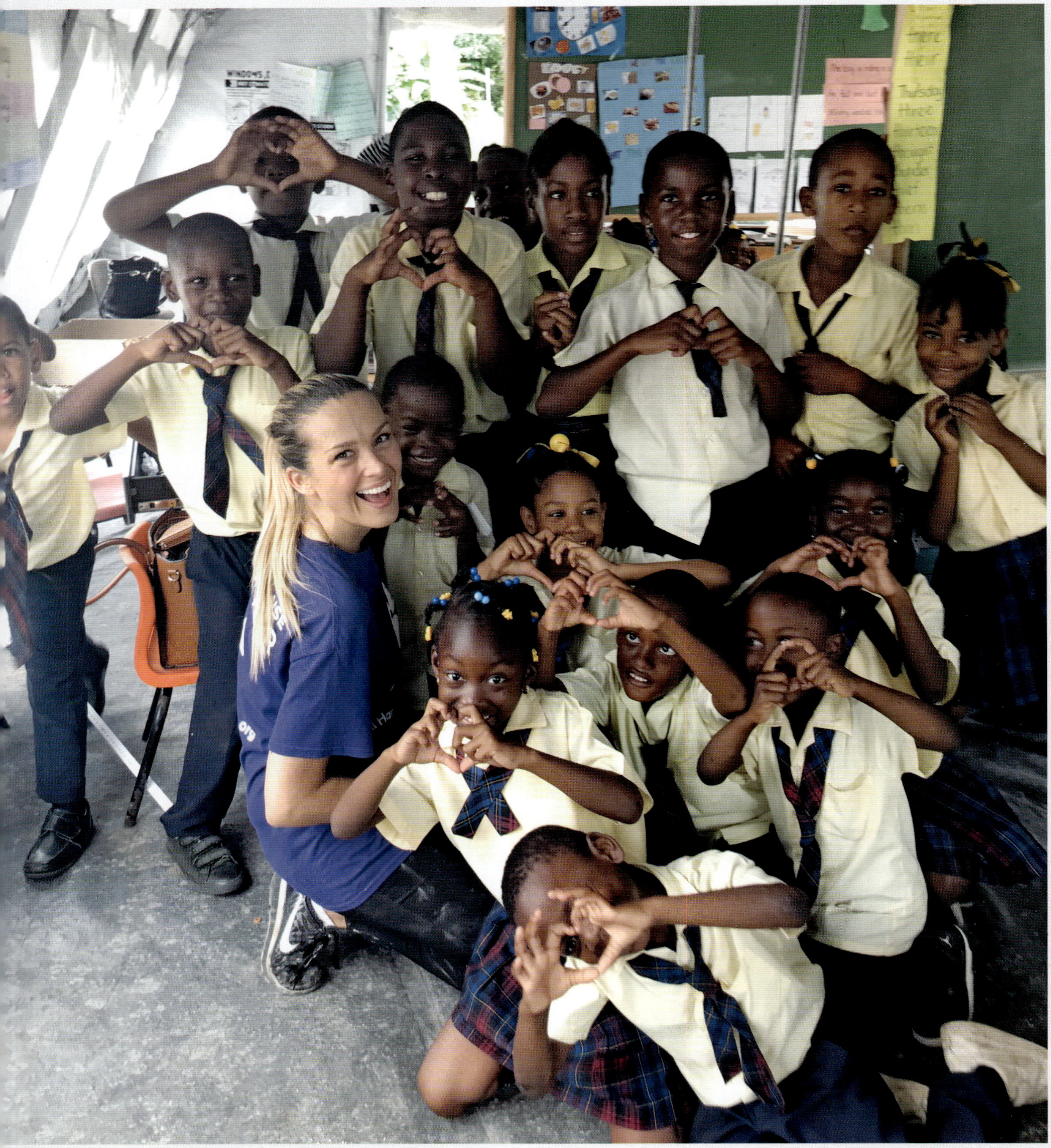

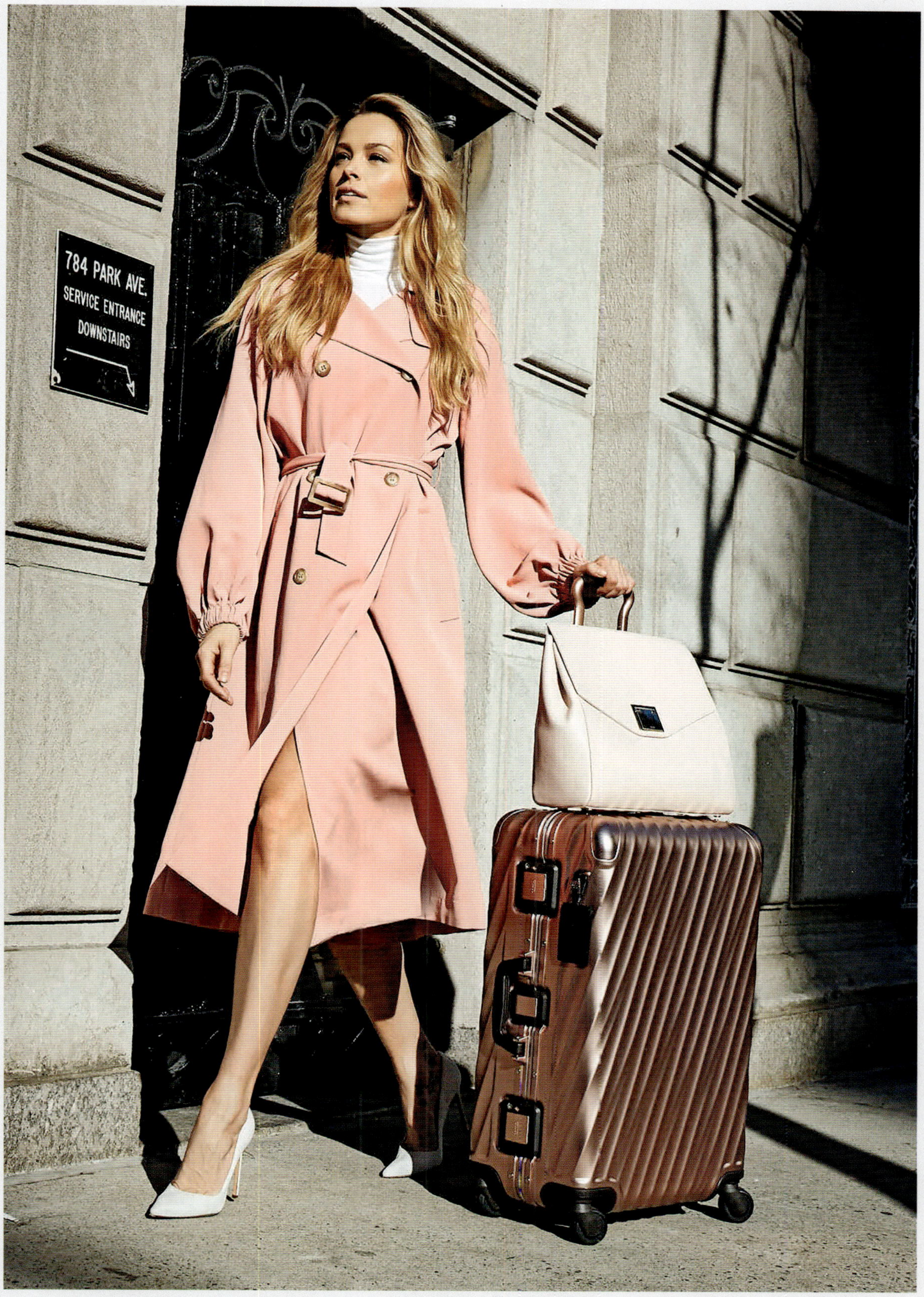
784 PARK AVE.
SERVICE ENTRANCE
DOWNSTAIRS

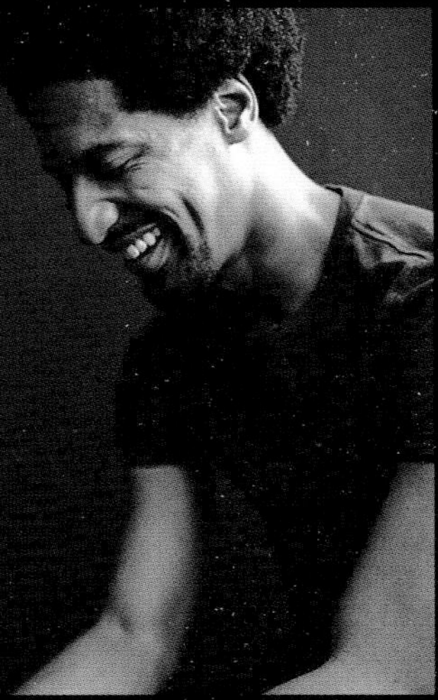

Collection

Alpha Bravo

Style

Cannon

Backpack

Collection

Launched

2014

Jon Batiste

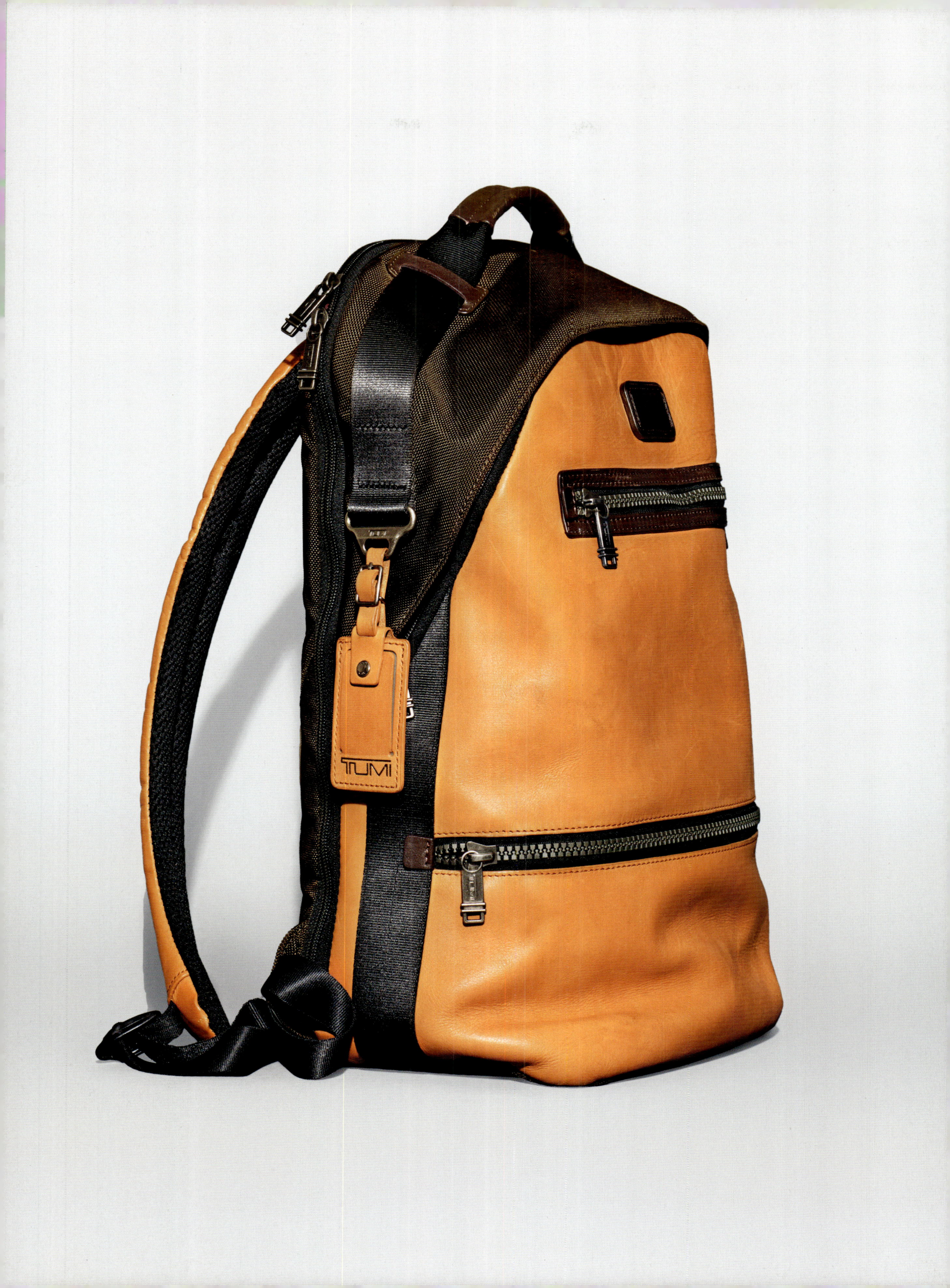
TUMI

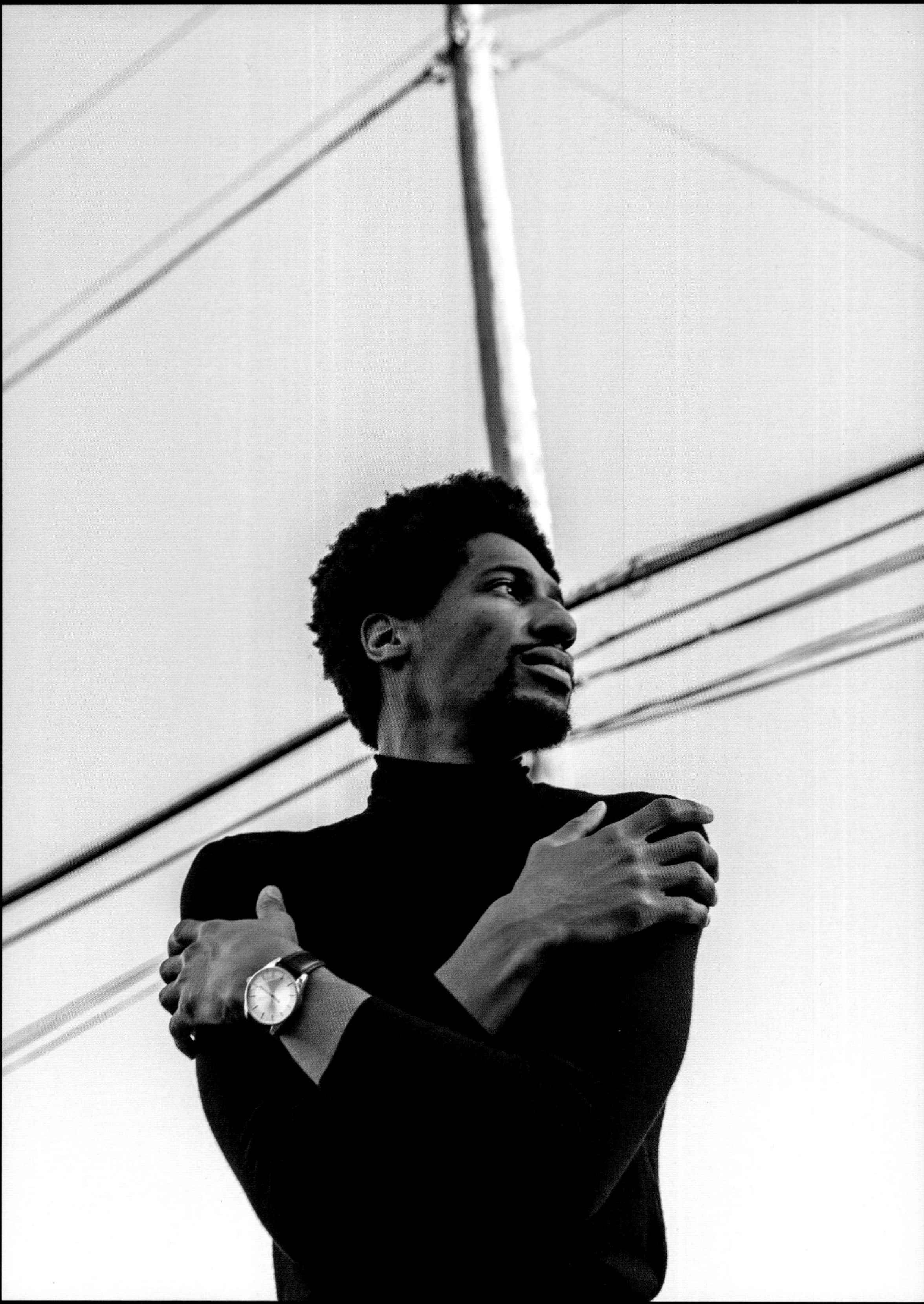

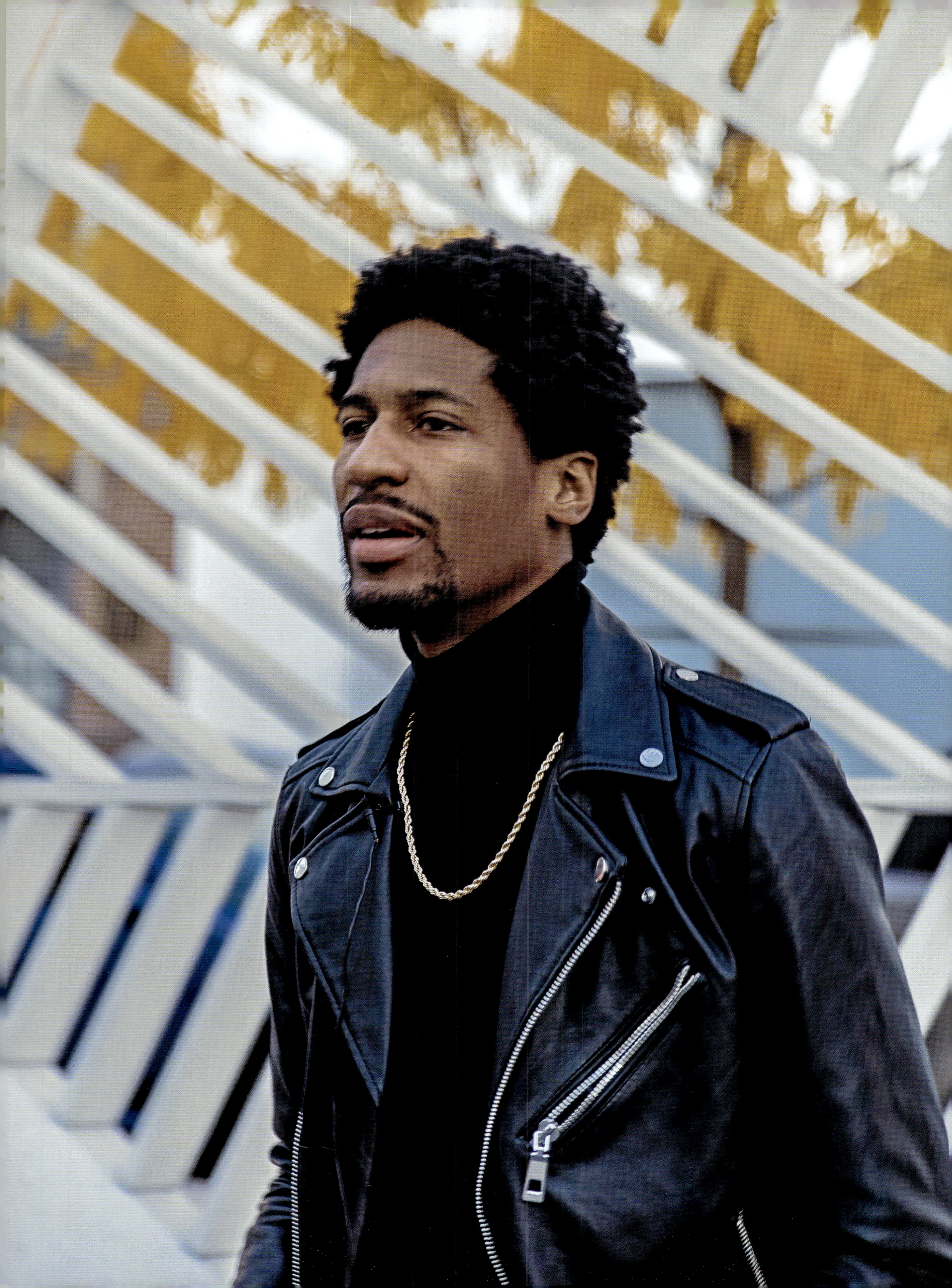

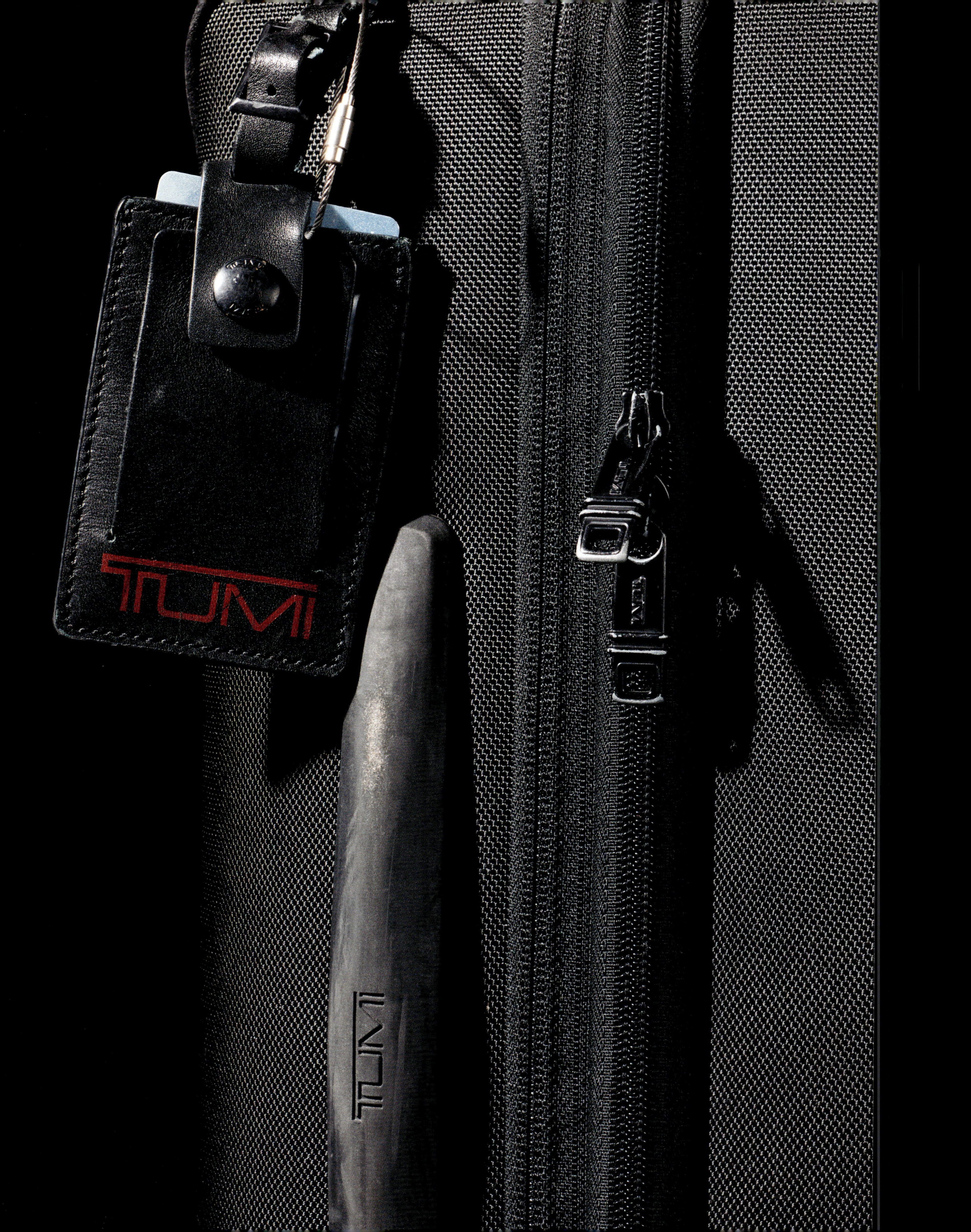
TUMI
TUMI

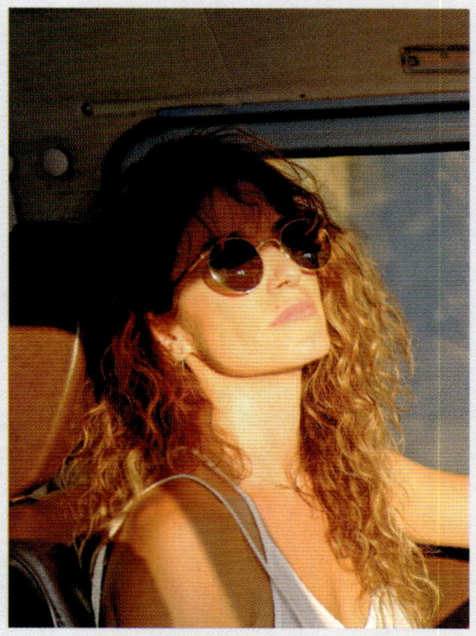

Nikki Reed

Entrepreneur, Activist, and Actress

Travel became my way of rebelling against all of the structure I had put in place for myself.

I didn't have a conventional childhood. From the time I was 13 years-old to about 18 years-old, I was working and around adults all the time, and so I became an adult pretty quickly. Then I watched my brother—who's only a year and a half older than me—turn 18 and move to Italy for art school, then down to Brazil to teach English. He had a traveler's spirit, and I soon realized that I couldn't do my job properly, portraying other people, if I didn't experience life in that same way. Before I started BaYou with Love, travel was my way of rebelling against all of the structure I had put in place for myself. I would hop on any plane going anywhere. I got used to packing in a carry-on. My TUMI carry-on is the perfect size. I know exactly where everything goes and how it'll fit. I know where my running shoes are supposed to go, and the compartment for my laptop and tablet—the compartments also make it really easy to see when you're forgetting something. I know I can pack everything in it—even for a three-week trip. It's my lucky travel bag.

Collection
Alpha 2

Style
Short Trip
Packing Case

Collection
Launched
2014

Russell Westbrook

Professional Basketball Player and Entrepreneur

I use the Alpha 2 hanging travel kit every time I travel. It just fits everything I need. My original one was a gift from my wife, Nina, and I carried it with me on the road. If you ever see a photo of me walking into an arena, I'm carrying that bag. One day, I was at a photo shoot and all of a sudden we smelled something burning in my dressing room. We started looking for what was burning and it turns out that when I had unzipped the travel kit, I had laid part of it on a lit candle and the bag was on fire. There was no serious damage to the room, luckily, but my beloved kit was ruined. So I went and got a new one the next day. Fast forward a couple of years and I was creating my own collection with them.

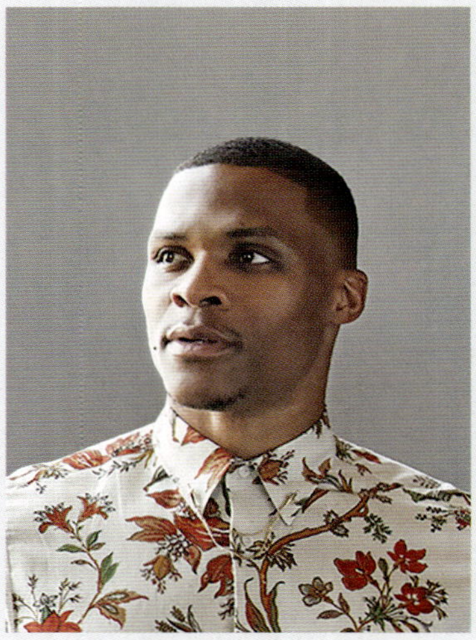

Collection
TUMI x Russell Westbrook

Style
Hanging Travel Kit

Collection Launched
2017

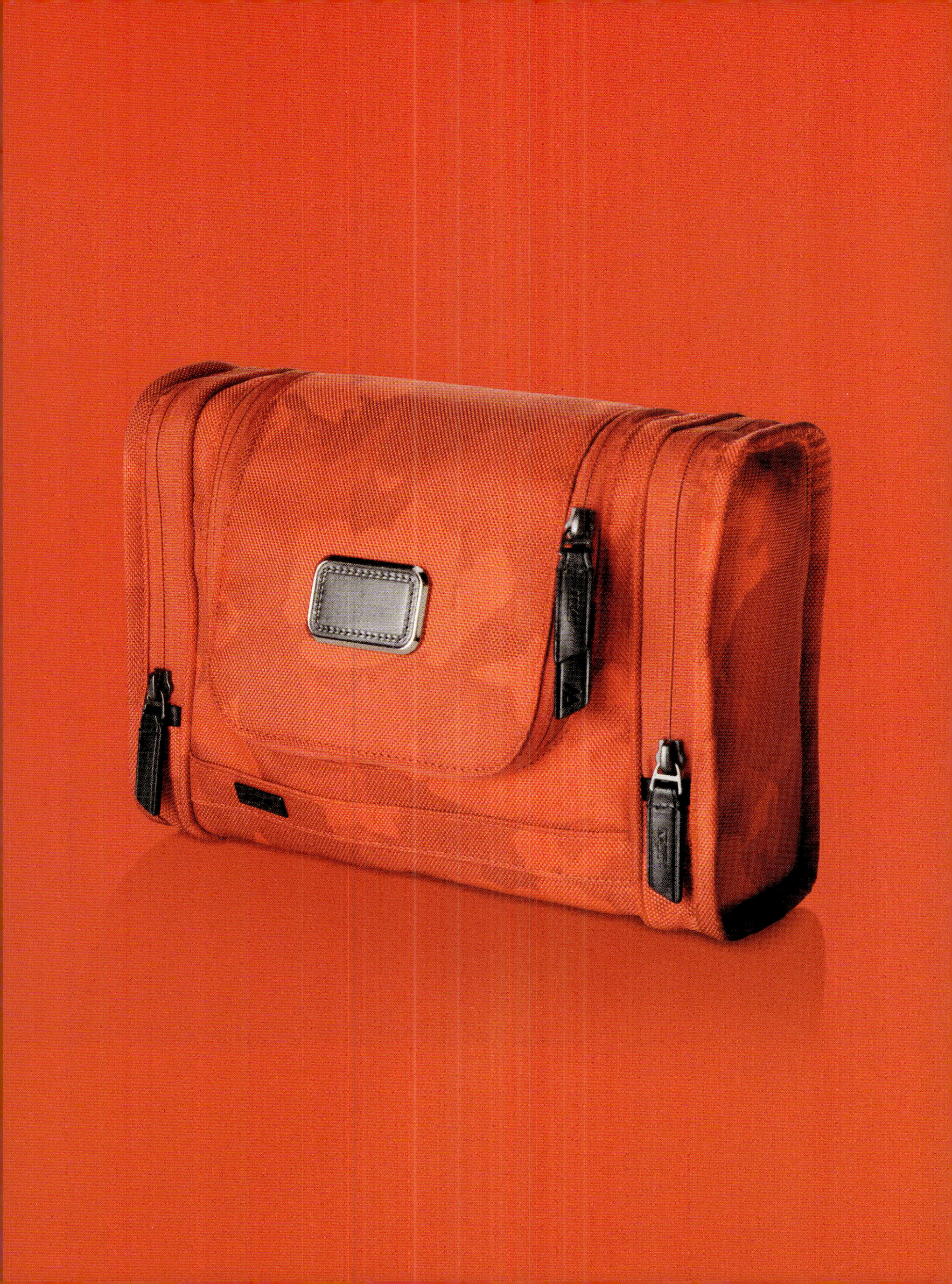

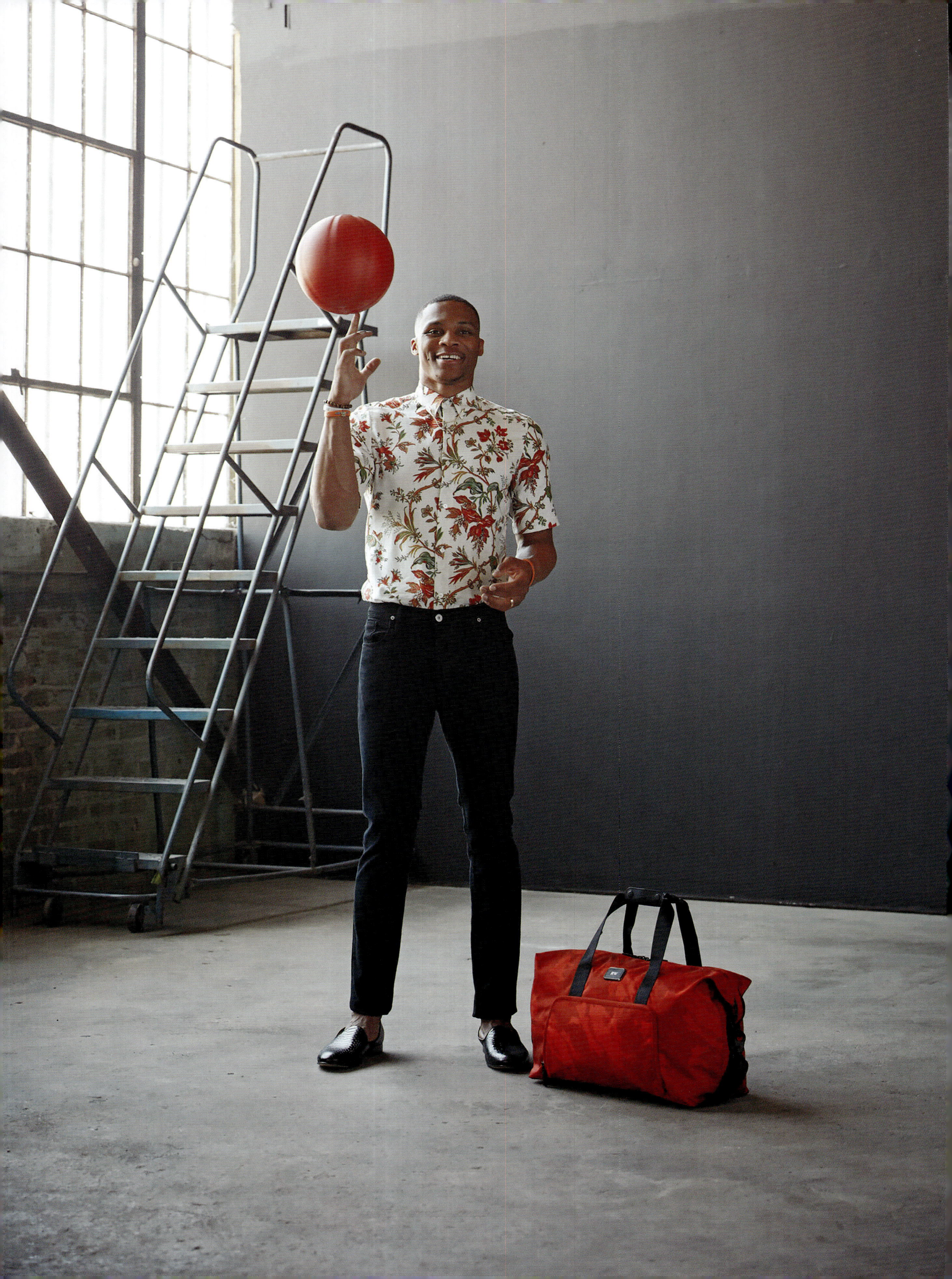

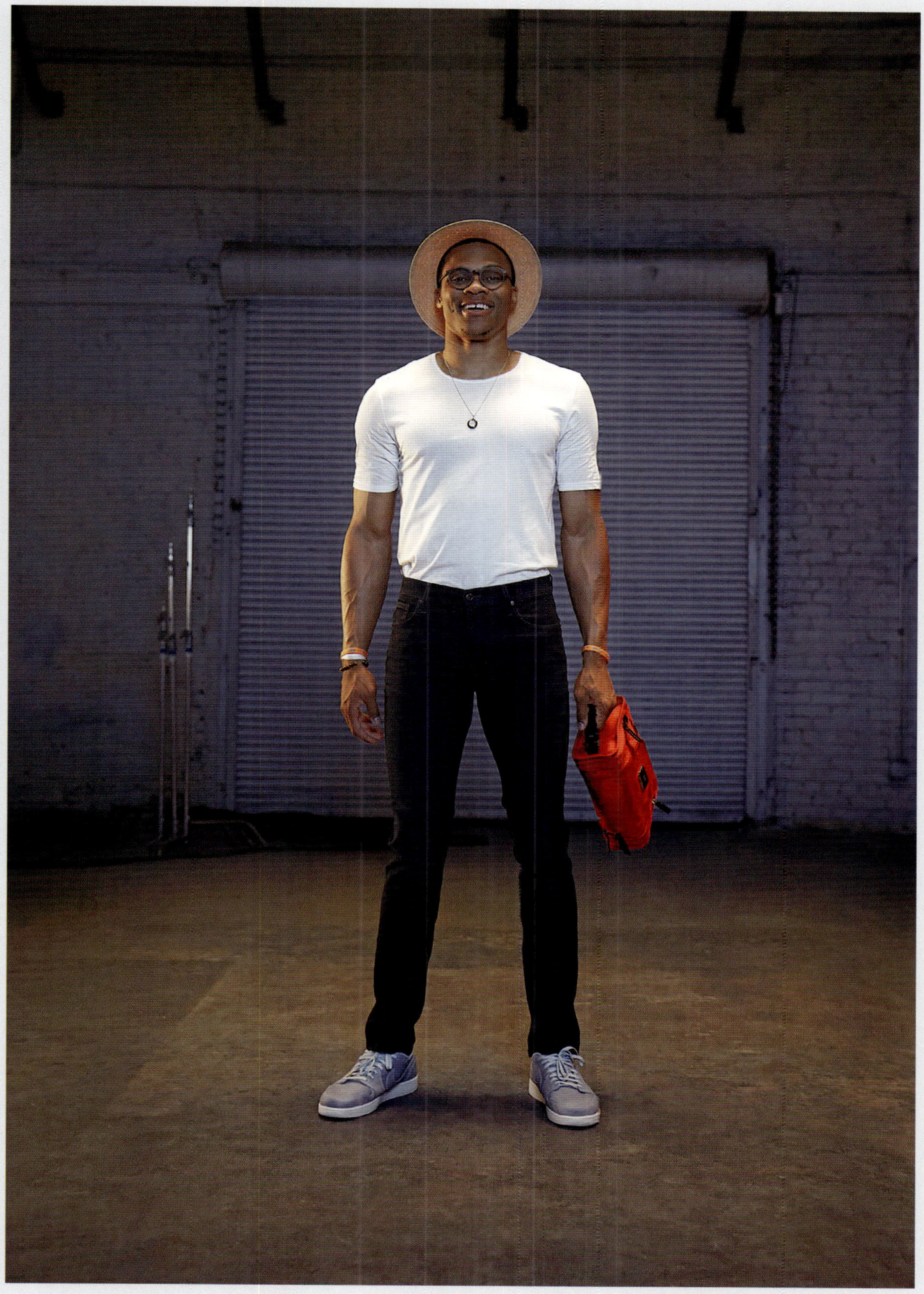

If you ever see a photo of me walking into an arena, I'm carrying that bag.

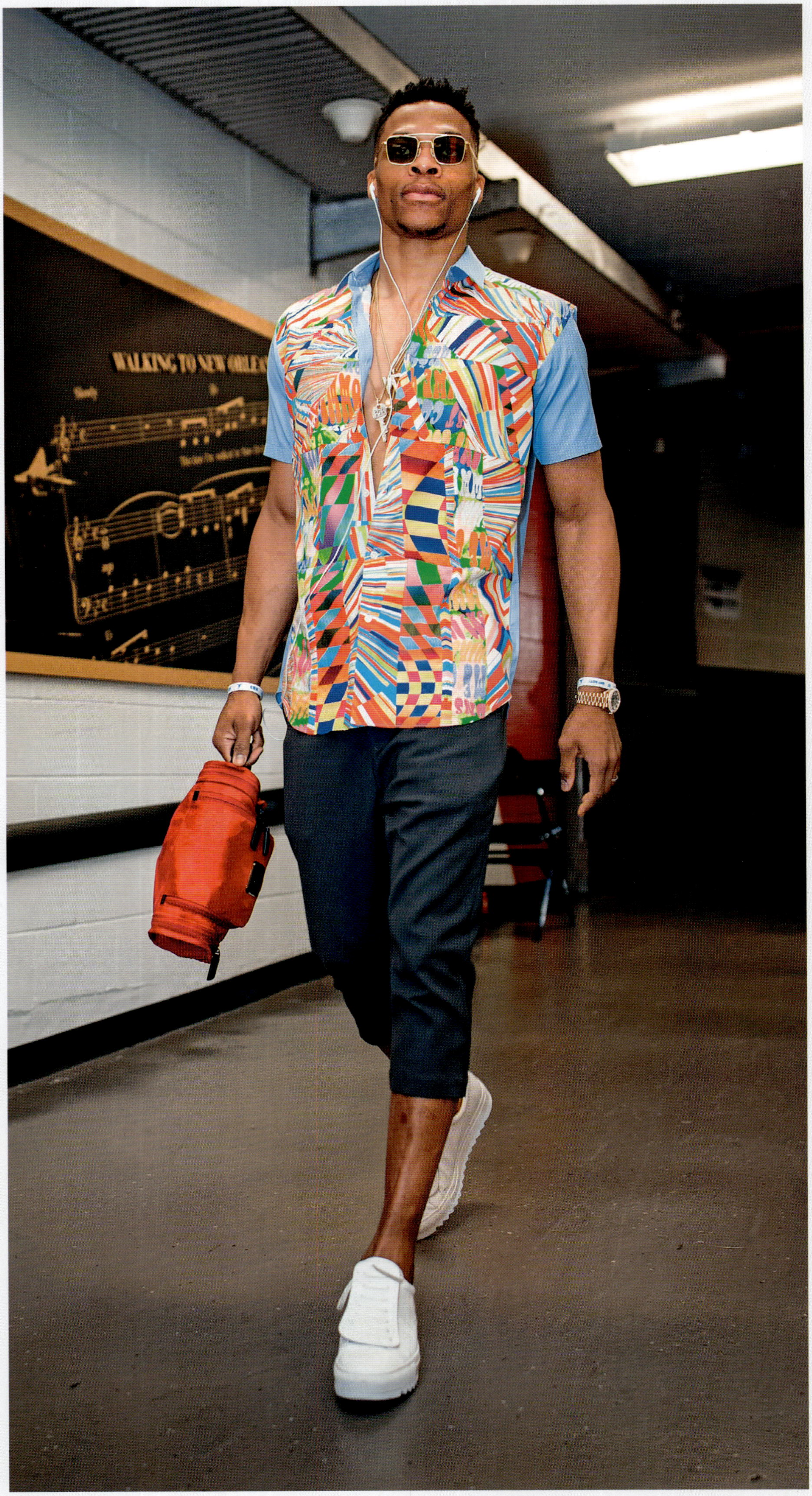

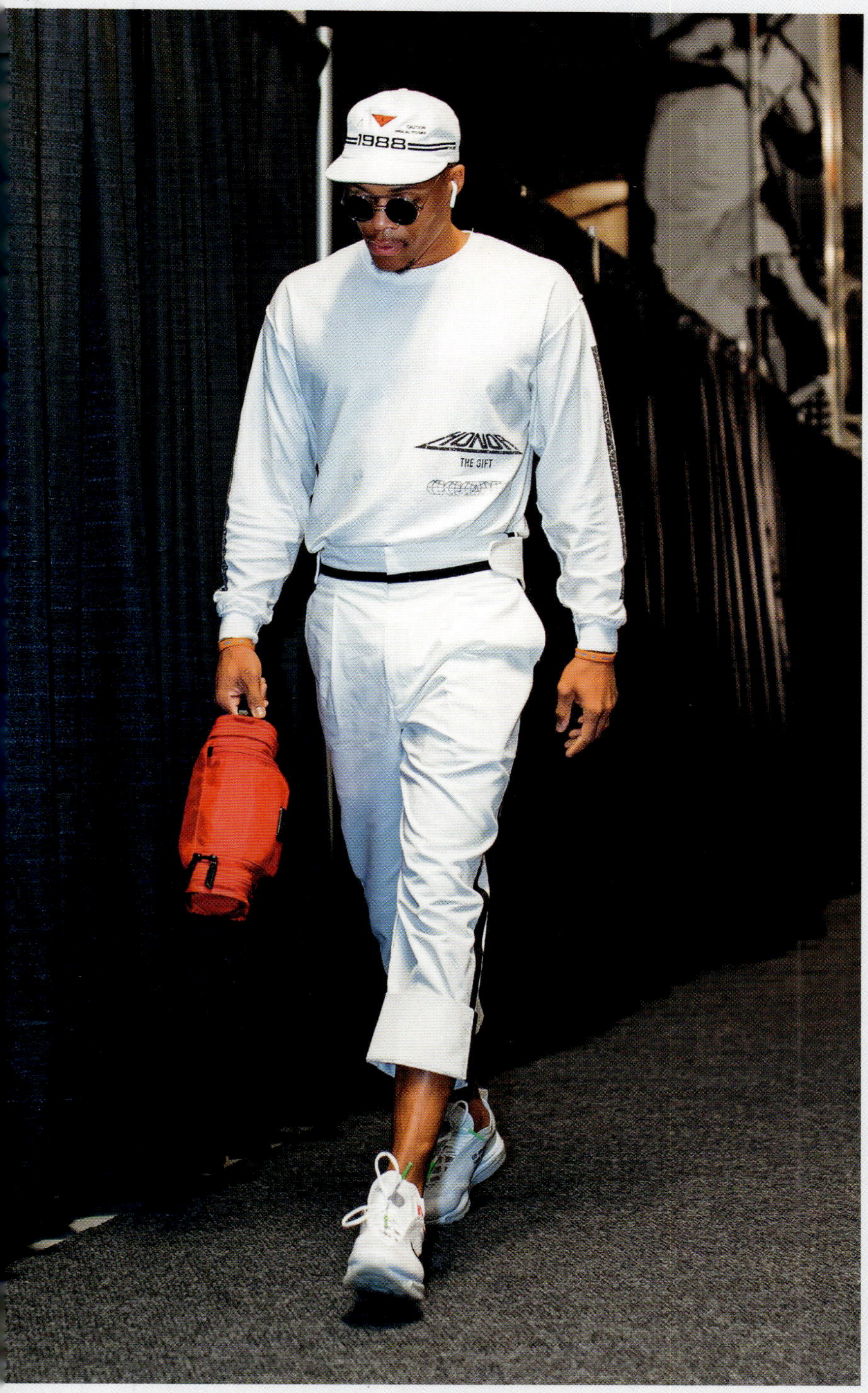
1988
THE GIFT

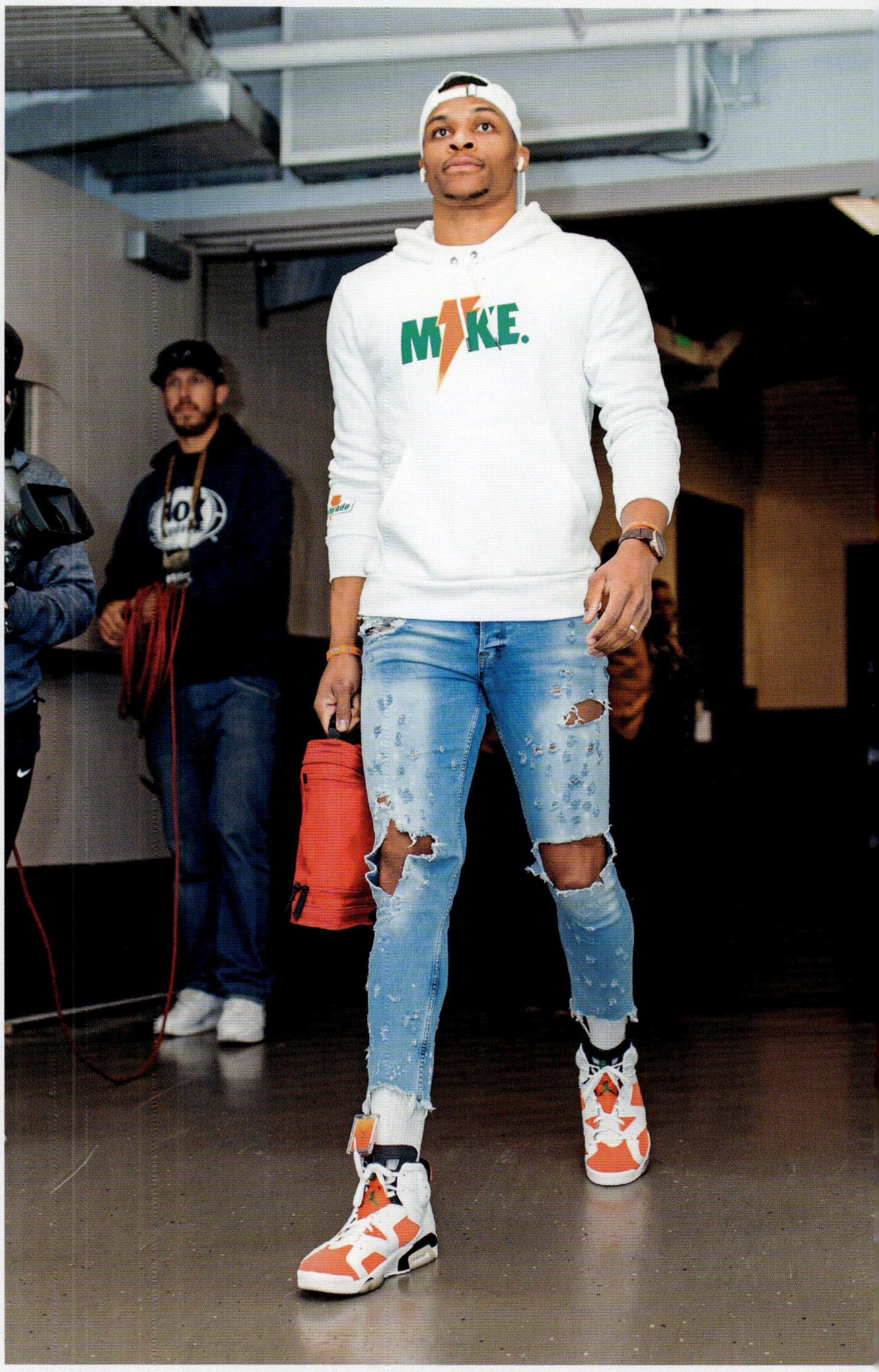

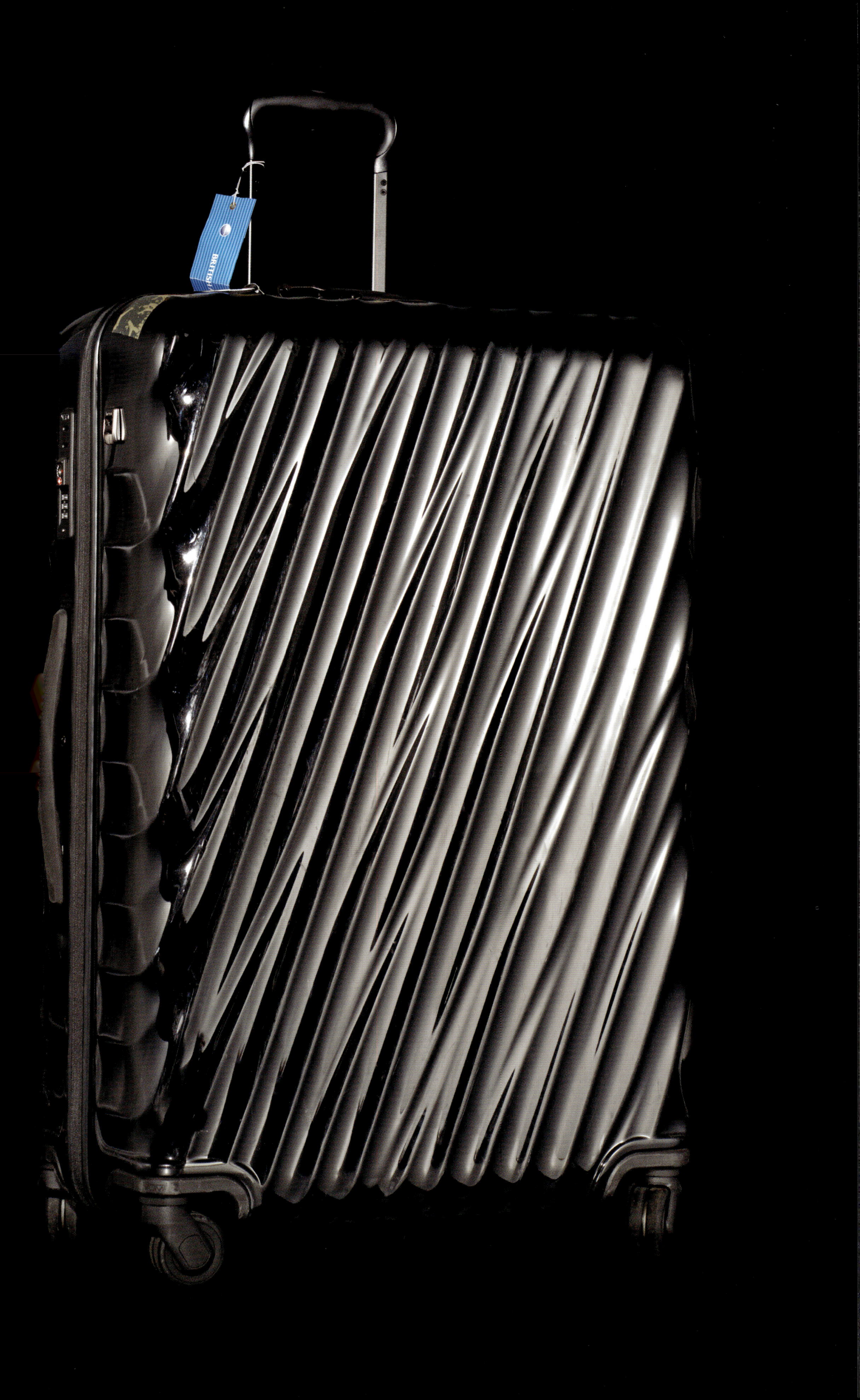

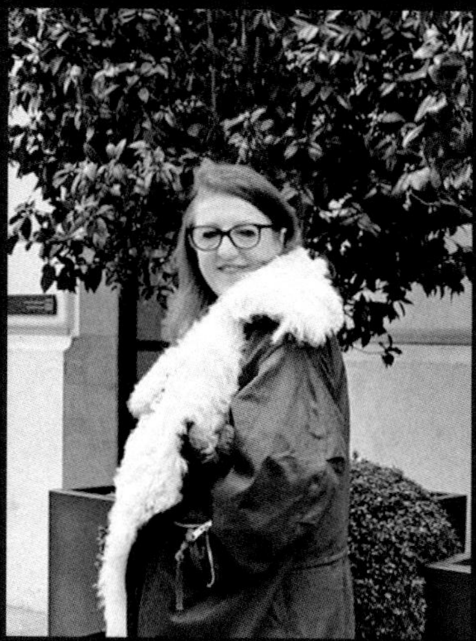

Collection

19 Degree PC

Style

Short Trip
Packing Case

Collection
Launched

2016

I'm not the world's lightest packer but I try not to carry more than I have to.

Glenda Bailey

Harper's Bazaar

I've known of TUMI for as long as I can remember. When I travel to Europe twice a year for fashion shows, I'm gone for three weeks at a time, so I need two big suitcases plus a soft carry-on. TUMI's hardshell cases are incredibly durable. I like to plan my outfits in advance. I sketch each look so I know exactly which accessories I'll need. I can usually get through most trips with three pairs of boots: knee-high boots, mid-calf boots, and flat boots. And then, of course, I need heels for events. As for bags, I have a top-handle that goes with practically everything. I always pack with acid-free tissue paper so I can save time steaming my things. Once many years ago I was sat opposite the Rolling Stones on the Concorde. Mick Jagger asked his bandmates if they'd join him for a drink as soon as they landed and Charlie Watts said, "Sorry mate, I've got to unpack." He should have used acid-free tissue paper!

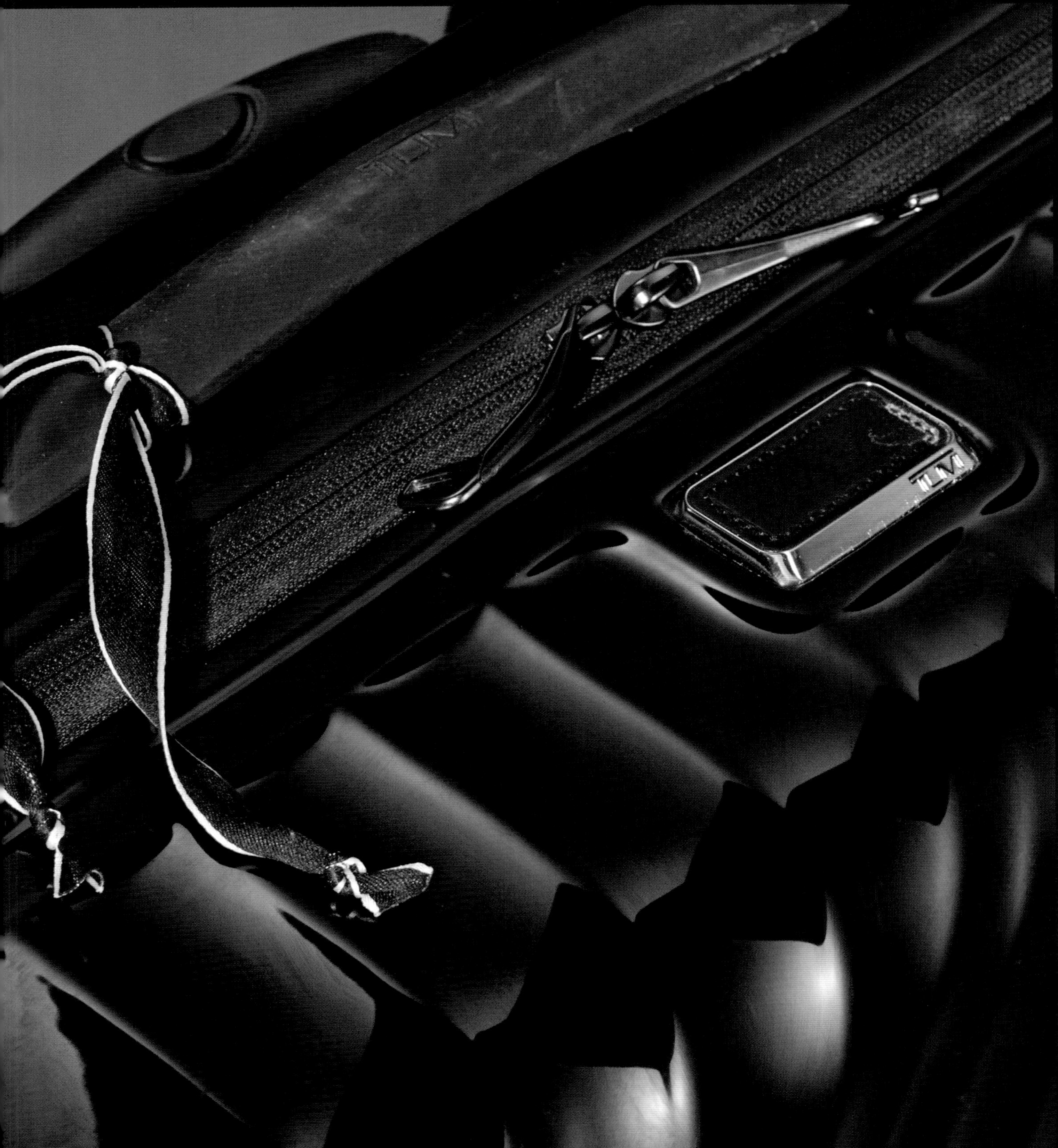

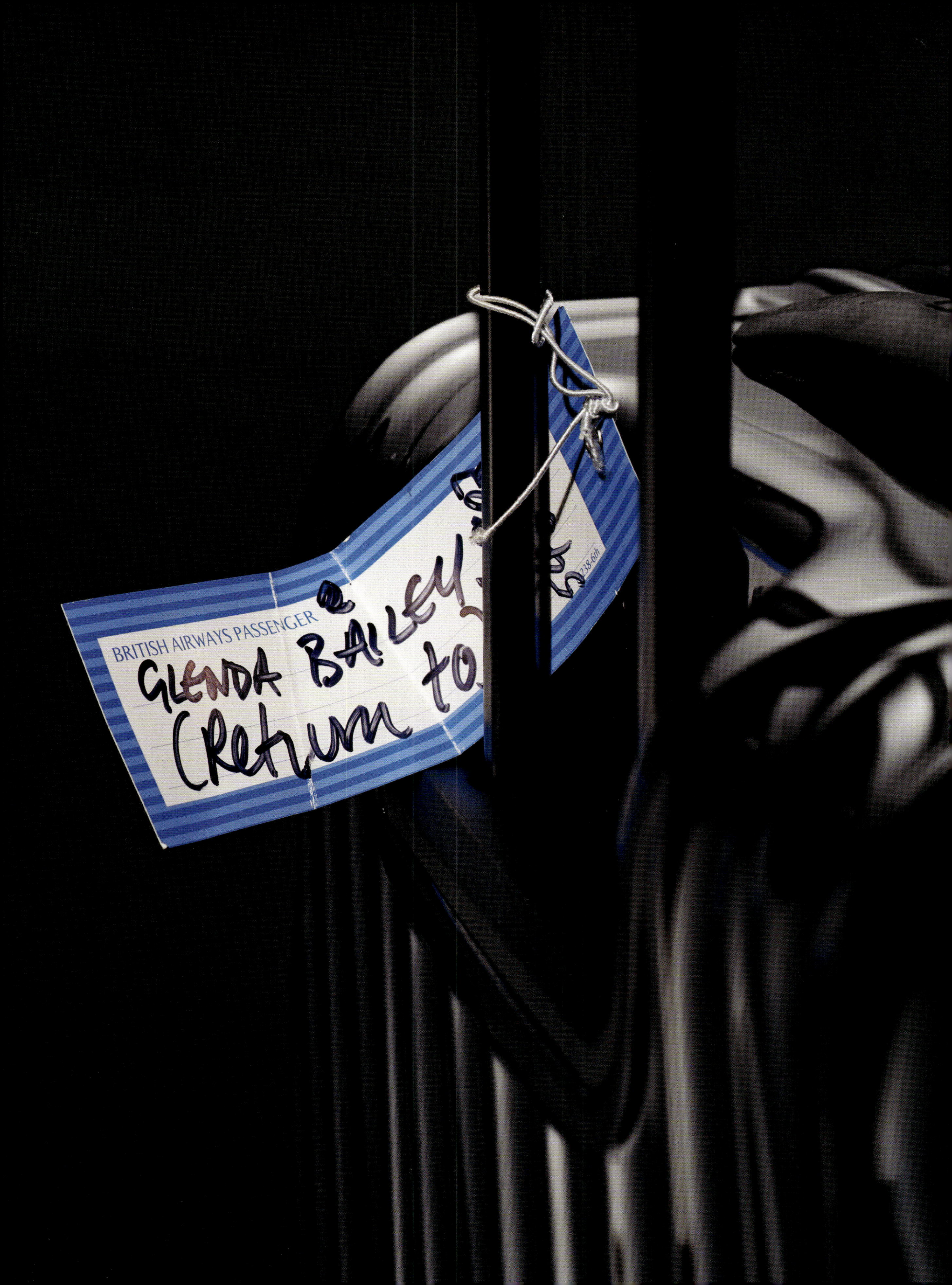
BRITISH AIRWAYS PASSENGER
GLENDA BAILEY
CReturn to

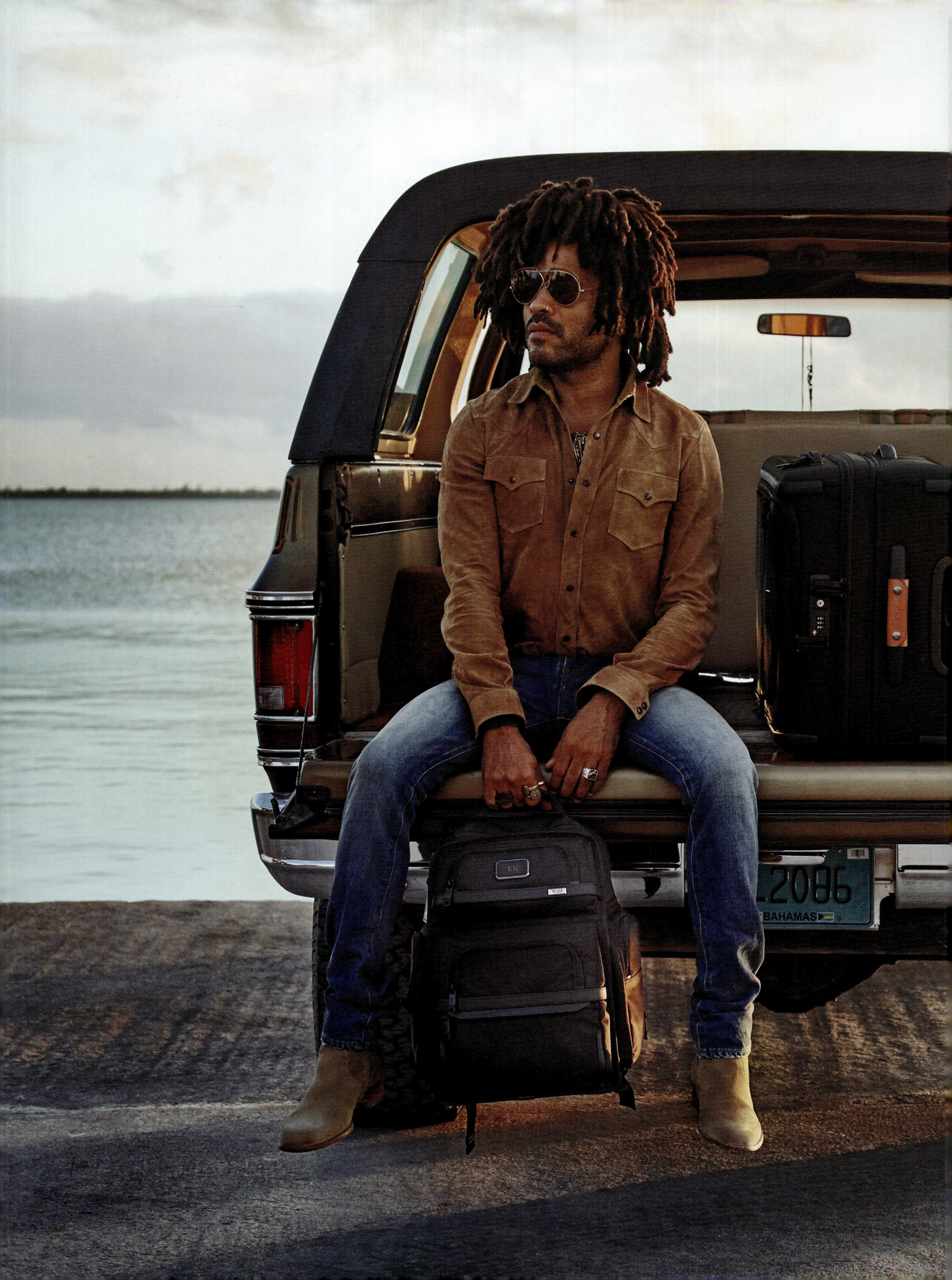
BAHAMAS

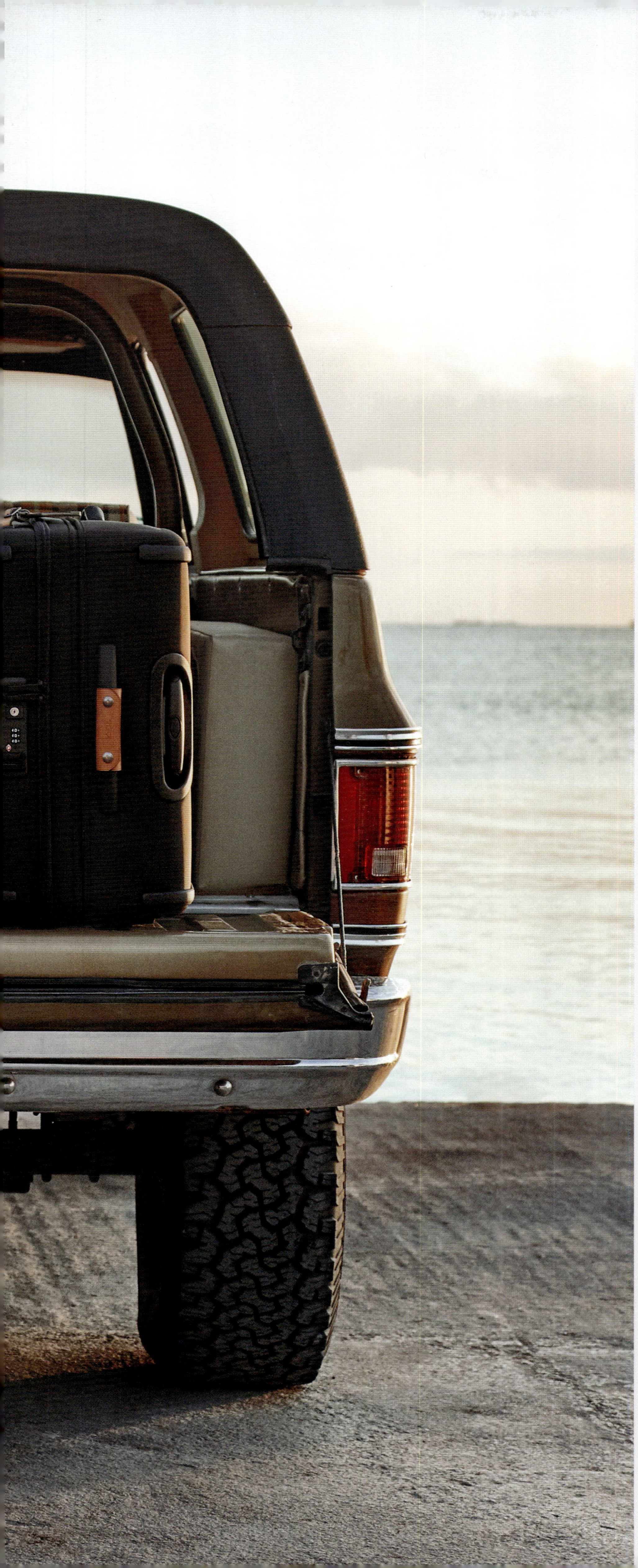

Lenny Kravitz

Music Legend, Designer, and Actor

I've been a TUMI customer and admirer of the company for as long as I've been touring. What attracted me to working with the brand is my love and passion for it. When you're traveling and living out of a suitcase, your luggage becomes an extension of your home—and yourself for that matter. It's the one thing that's always with you: on the plane, on the tour bus, in the hotel room, backstage, or on set. So, you want it to be reliable, durable, and make a statement. For me, TUMI embodies all of those qualities to the fullest. Working on the campaign was very gratifying. Not only did I get the opportunity to represent a great product but it was a very collaborative experience as I was involved in many creative aspects of the campaign.

Collection

Alpha 3

Style

International
Carry-On

Collection
Launched

2019

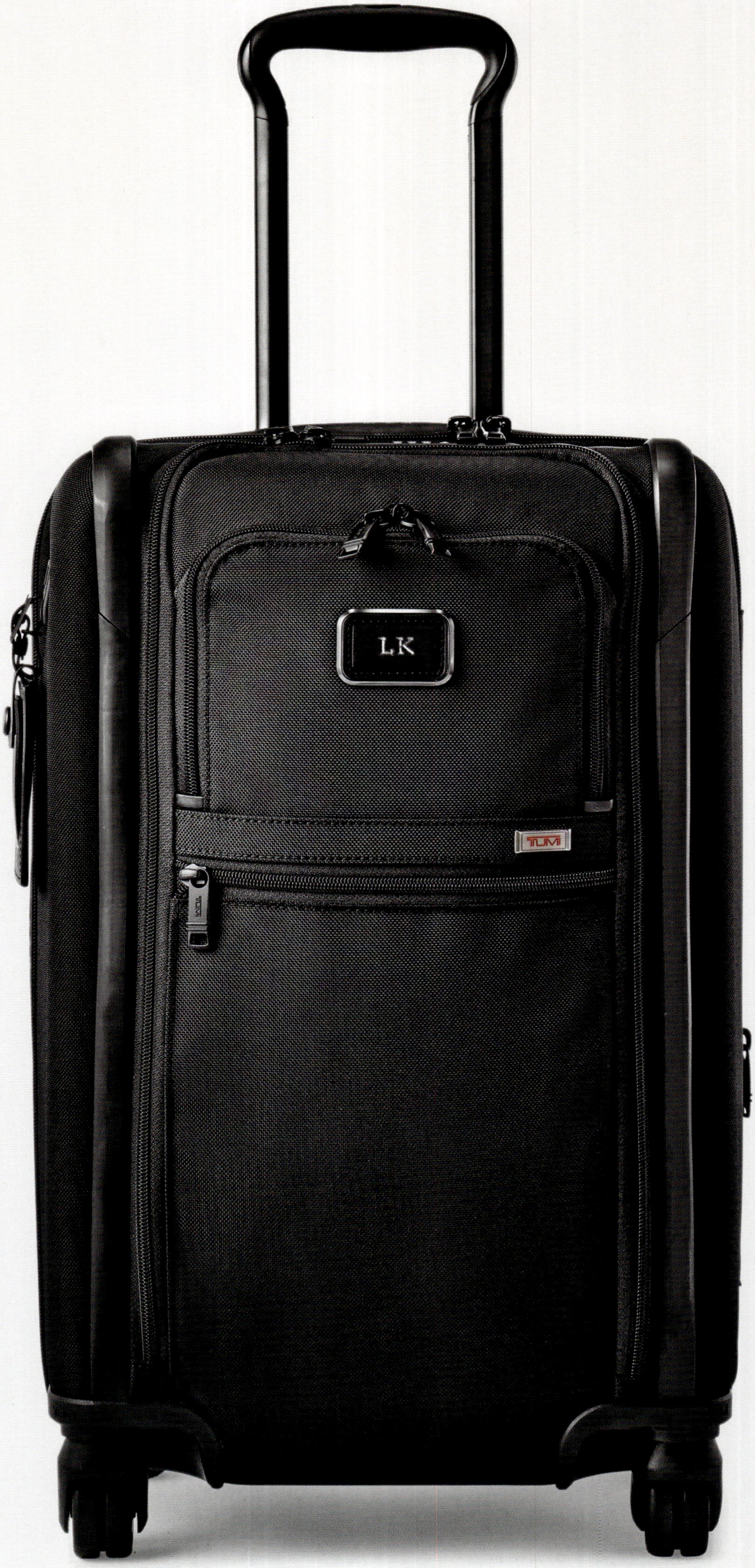

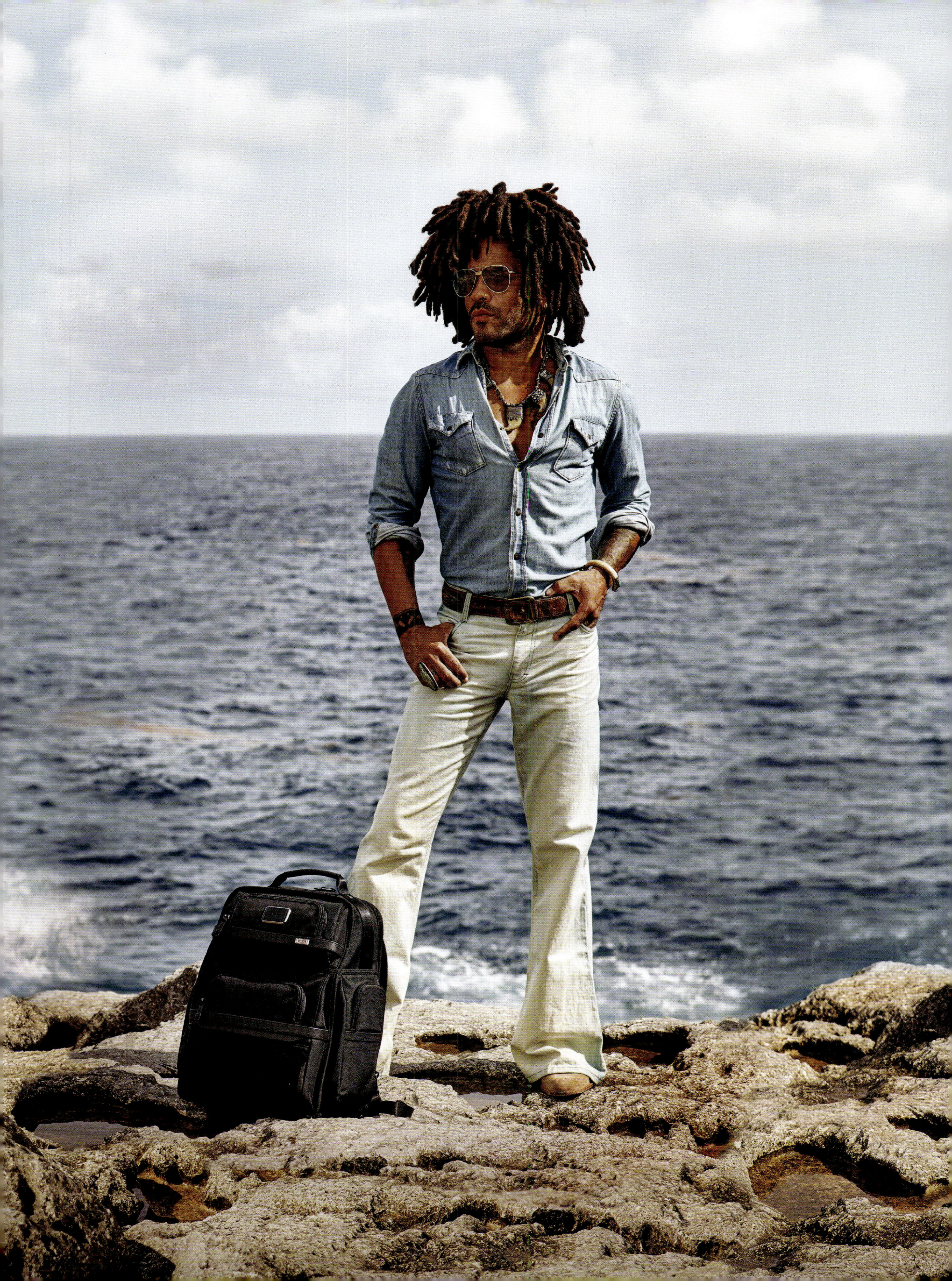

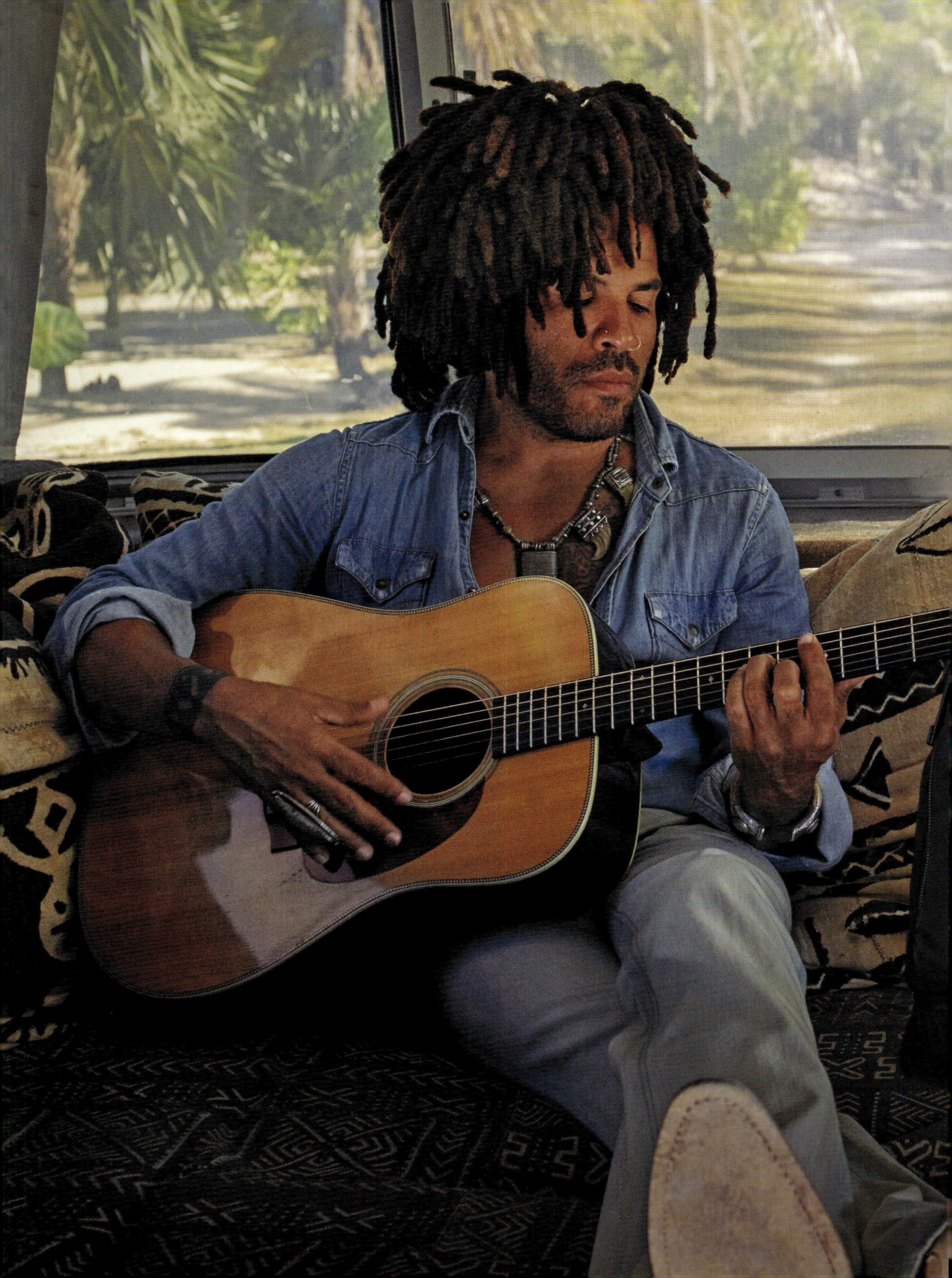

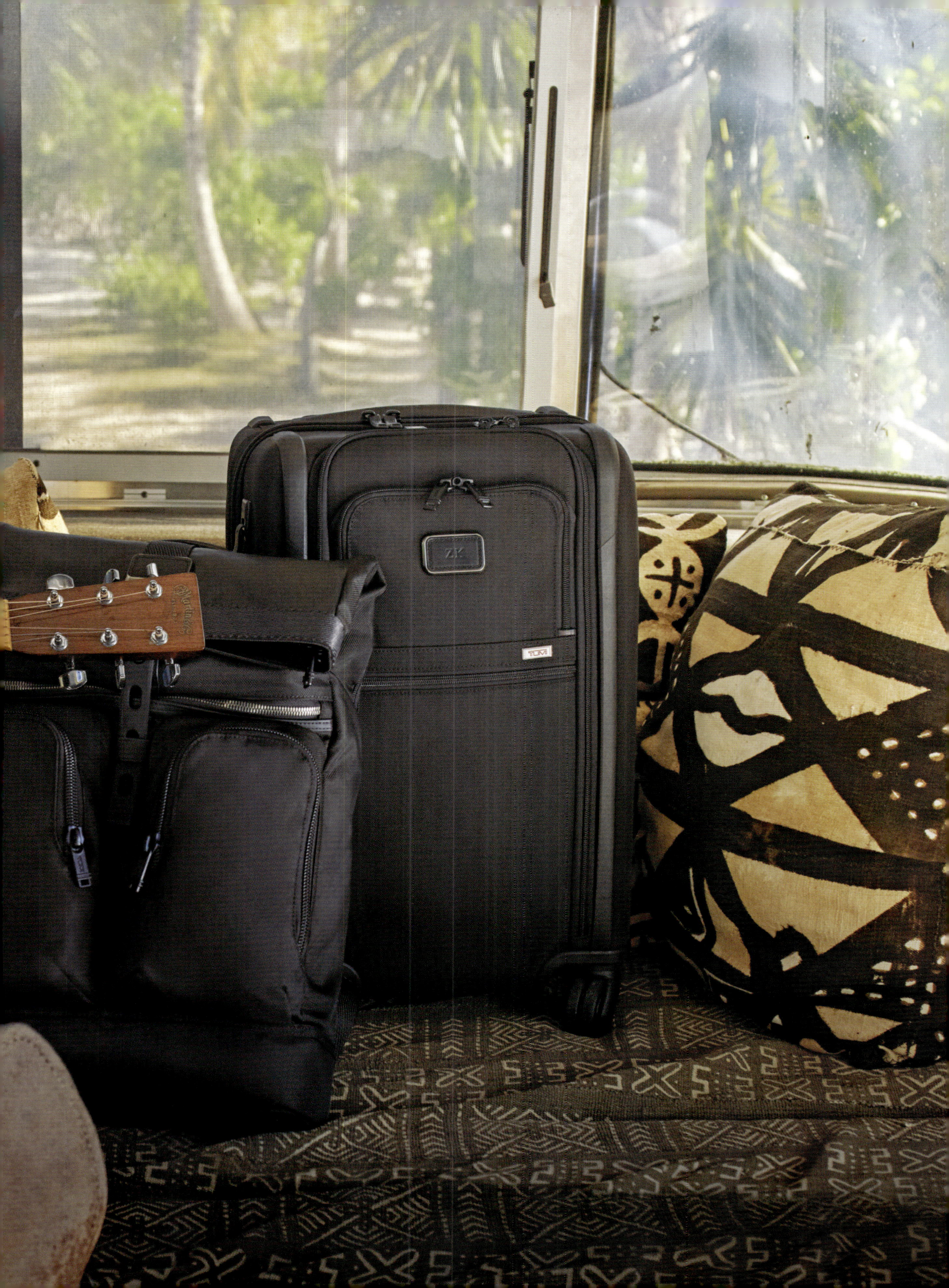
TUMI

Dillon Francis

Musician, Producer, and DJ

Skrillex introduced me to the brand, and now it's the only one I'll use.

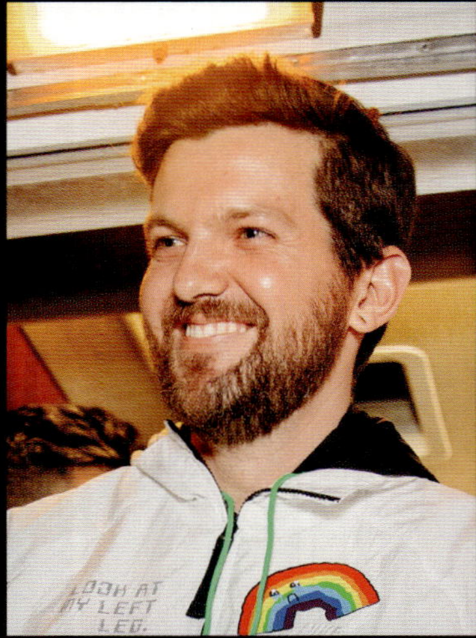

I've always known about TUMI, but I didn't start using it until I became a well-versed traveler. My friend, the DJ Skrillex has the leather backpack, the same one I have. I know where everything is in that bag: For instance, right now I know I have my Tide pen, digestive pills, pens, Apple air pods, and chapstick in one pocket. In another, I have my pre-made soup packets—I don't ever want to eat food on the plane. I can also fit two computers, power converters, chargers and all my clothes. I had to replace my backpack after four years. Not because it was broken—there was absolutely nothing wrong with it. The problem was, I spilled tequila all over it, and I couldn't get the smell out. It didn't make me change my drink of choice. But it did make me rethink where I placed my drink of choice.

Collection

Alpha 2

Style

Brief Pack

Collection Launched

2019

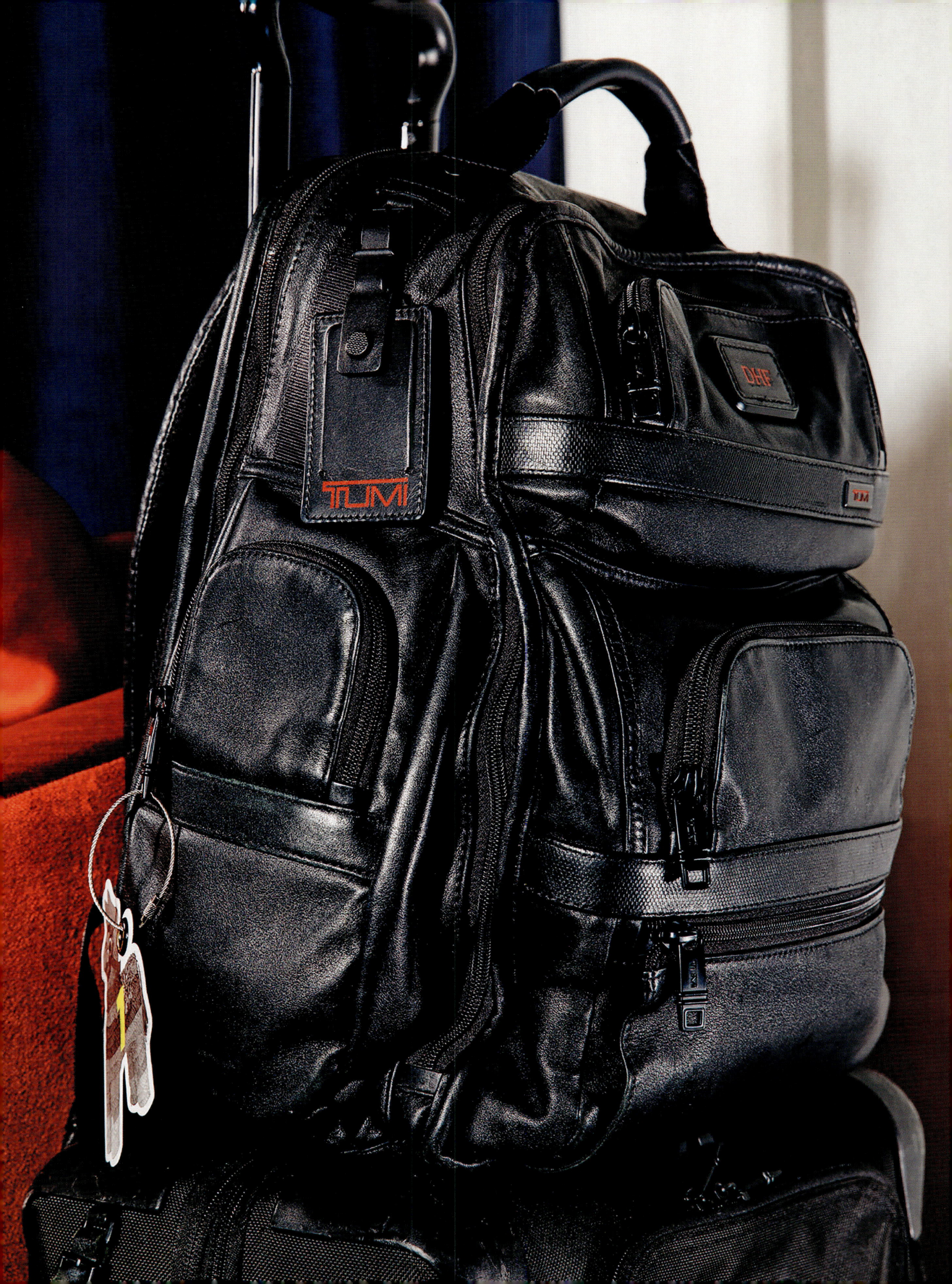
TUMI
DHF

Collection

Alpha 2

Style

Travel

Satchel

Collection

Launched

2016

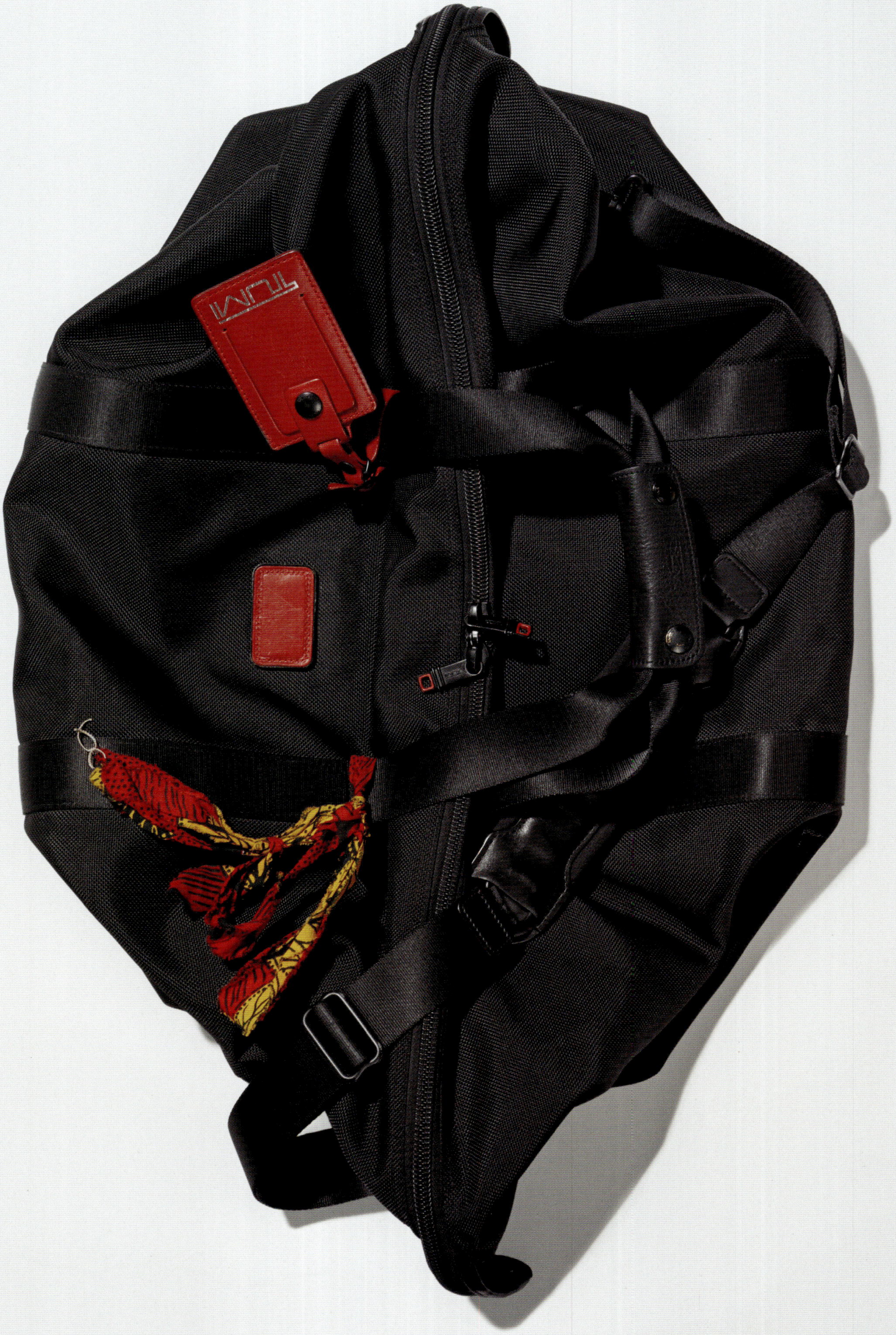

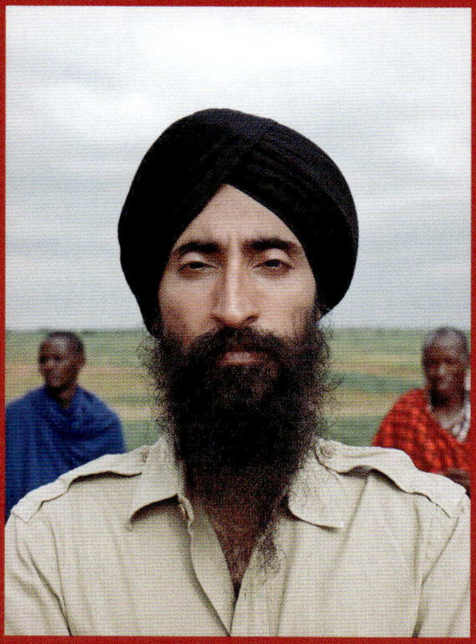

The simplicity of the materials—the handle, how the zippers work—makes me happy.

Waris Ahluwalia

Designer and Actor

TUMI has always been very functional. But then they evolved. The design changed. Once they did that, it was a home run for me. My black TUMI duffel is my favorite. It's so light, so soft. I get a kick out of physically carrying on. In fact, I can't remember the last time I rolled on board. I always pack at least one suit, even if I'm not planning on going anywhere that requires one. I tend to find myself in situations where they're required: in the fall or winter, I do dark blue or black. In the summertime, I go with something lighter. My TUMI pieces—I also have the backpack—are functional and pleasing. I've used my duffel on so many trips and it shows barely any wear and tear. This bag has been there for me.

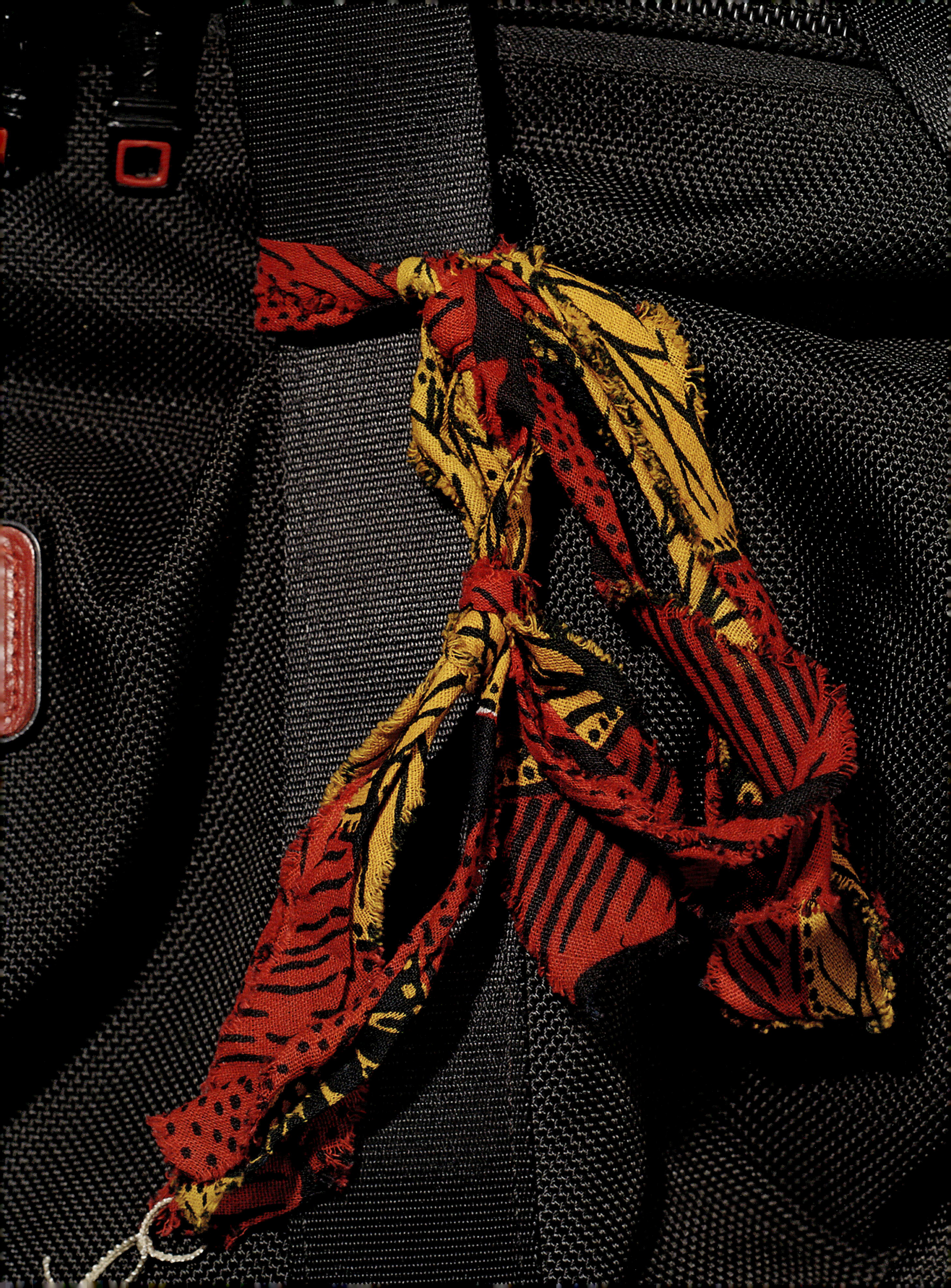

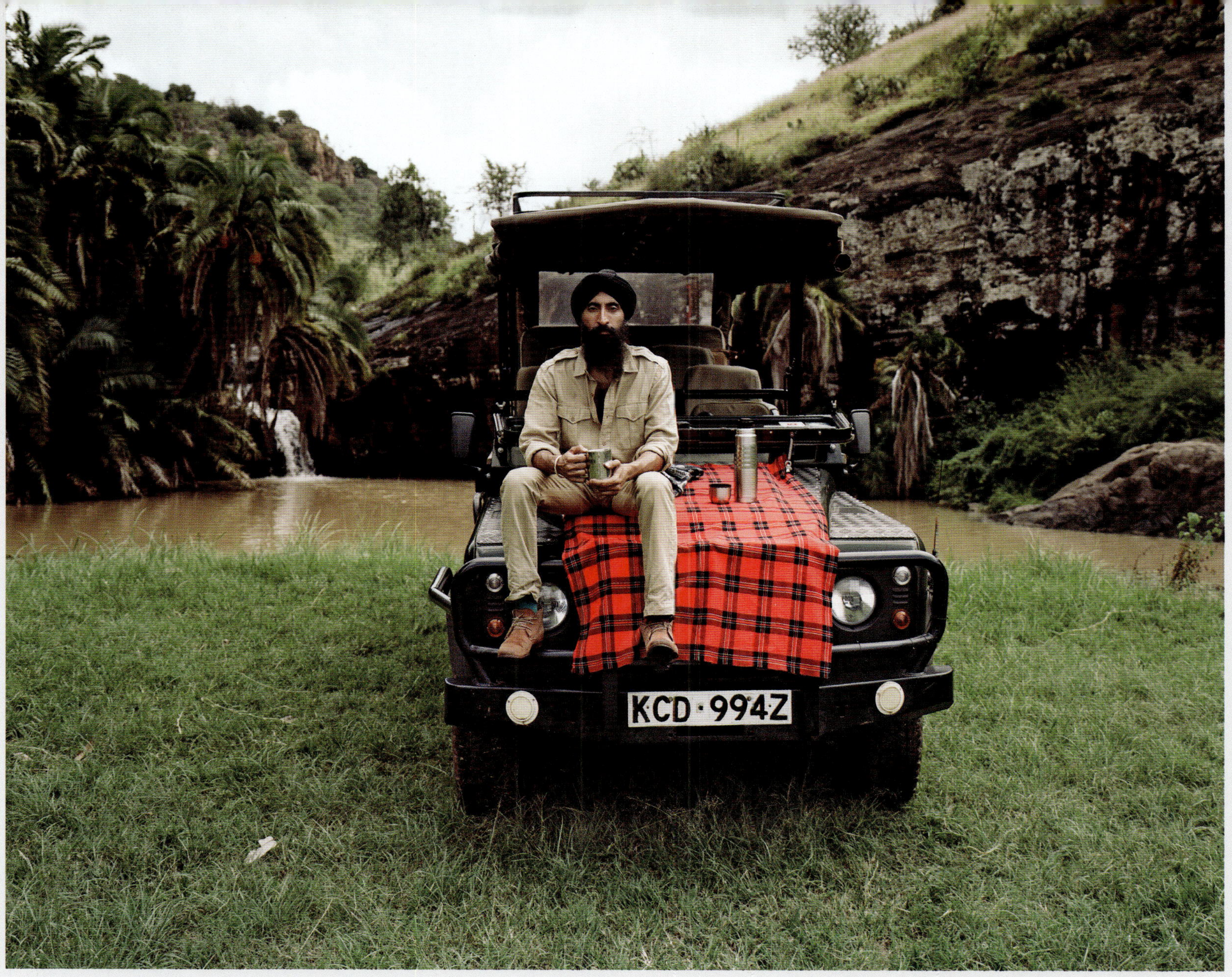
KCD·994Z

TUMI bags are meant to be functional for so many people, but they work so specifically for what we do.

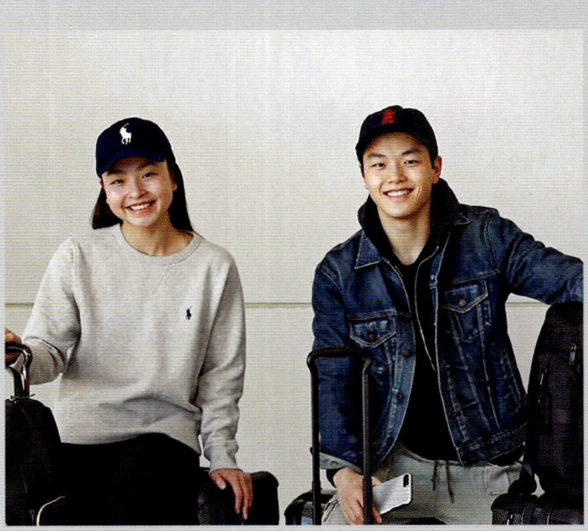

Maia and Alex Shibutani

Siblings and Two-time Olympic Medalists

AS: Our dad introduced us to the brand. He'd always use TUMI on work trips, so, naturally, we used them when we started skating. We needed a durable suitcase—if our blades get nicked or damaged, it would affect our ability to be at peak performance. We're always confident nothing will happen to them while they're in our luggage.

MS: We brought our skates to and from the rink every day when we were training for the Olympics. We'd pack them in our International Carry-On. People would think we were going on trips, but really it's the best way to protect our equipment. And it's perfect as is. TUMI bags are meant to be functional for so many people, but they work so specifically for what we do. We keep our skates in one half and our costumes in the back half so, when we travel, all the important things we need for competition are with us at all times. And with the two-carry-on rule, we both rock a backpack so we can carry on as much as possible.

LEFT

Collection

Custom CFX

Style

International Carry-On

Collection Launched

2018

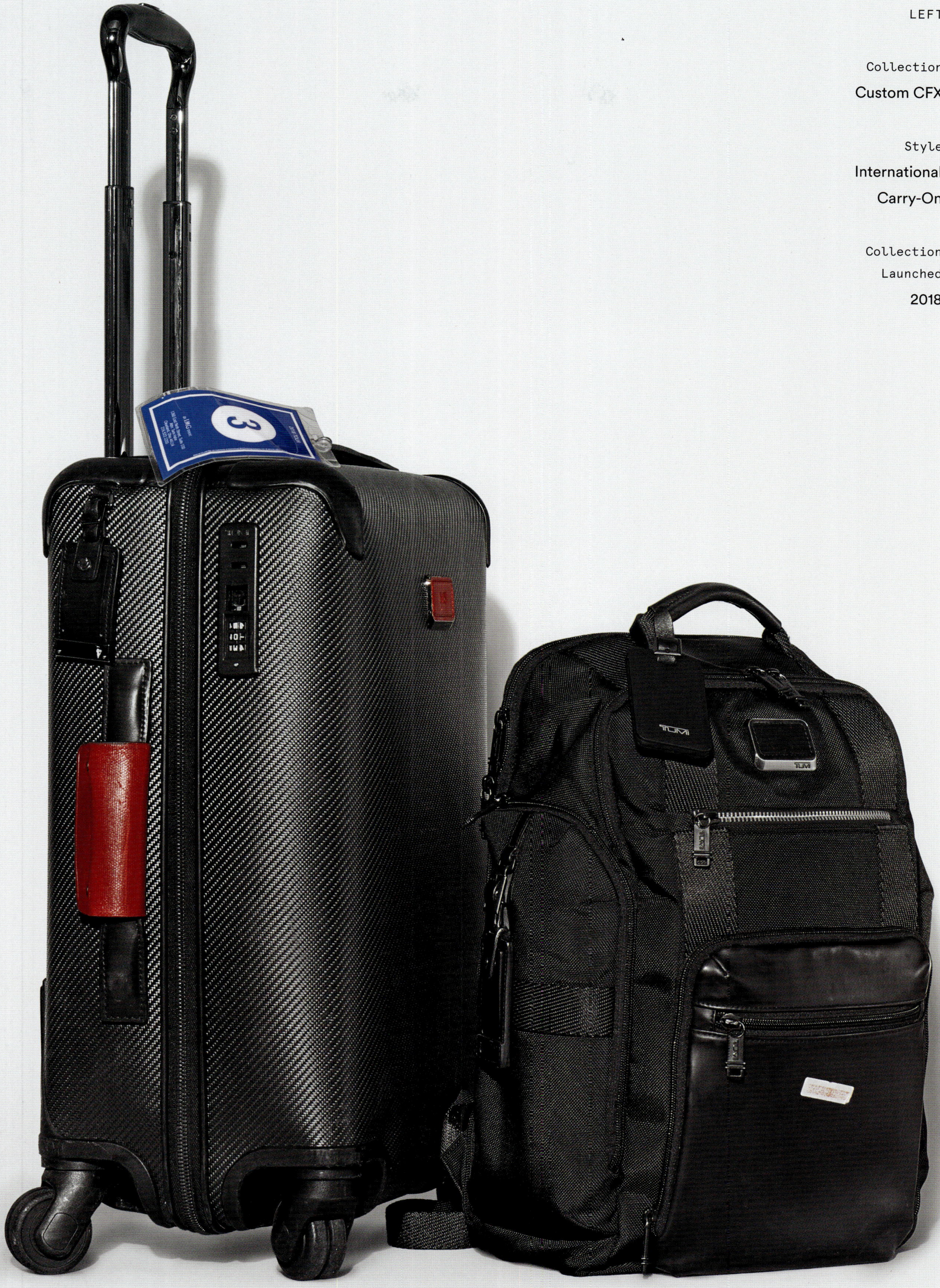

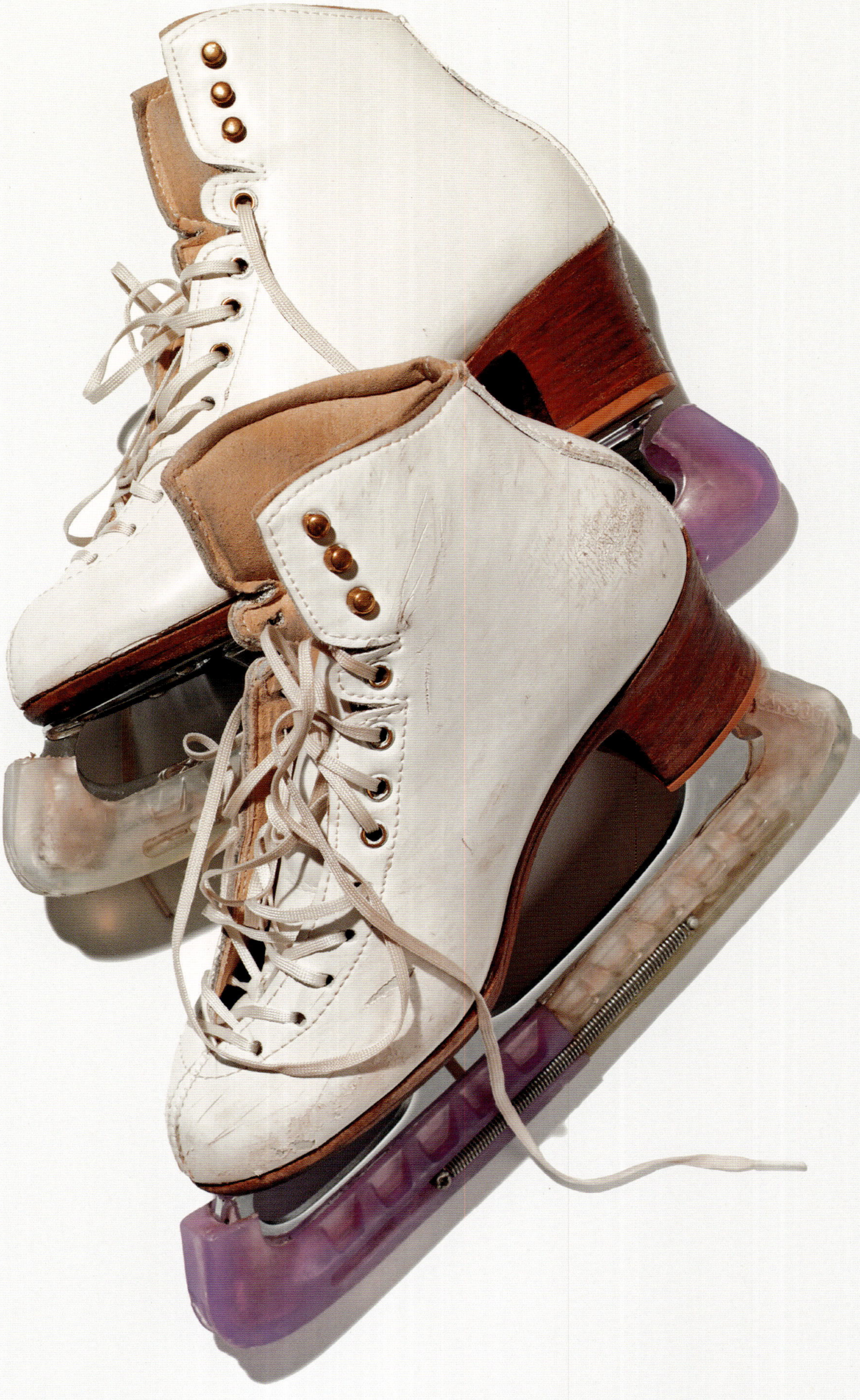

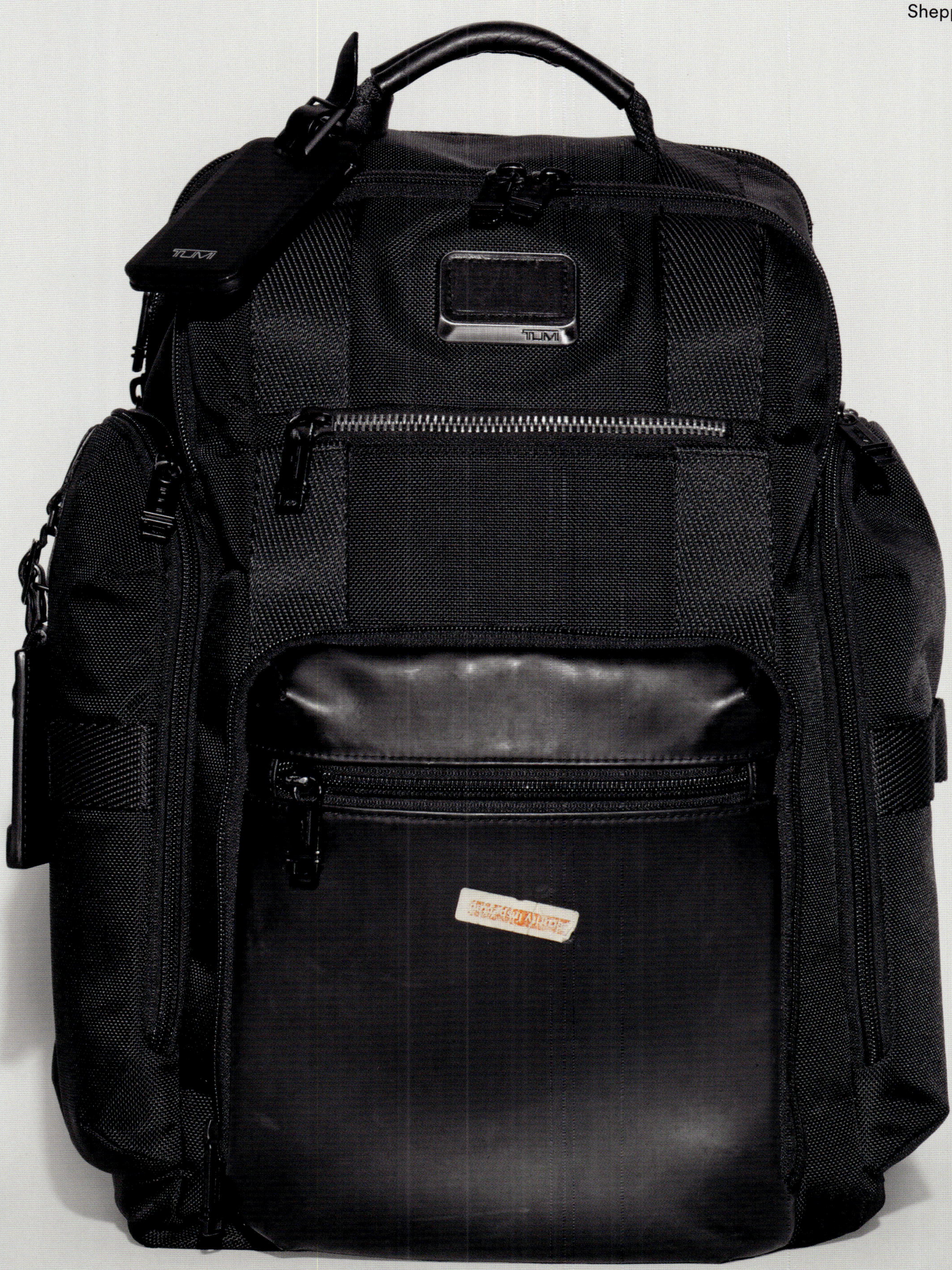

Collection

Alpha Bravo

Style

Sheppard Deluxe
Brief Pack

Collection
Launched

2018

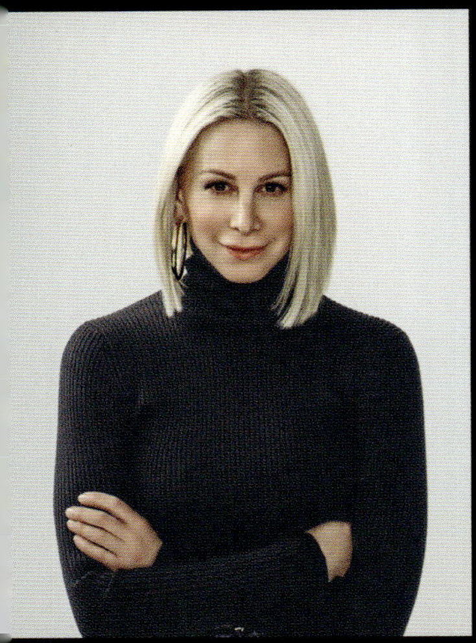

Jennifer Fisher

Founder and Creative Director, Jennifer Fisher Jewelry

Collection
19 Degree Aluminum

Style
Short Trip Packing Case

Collection Launched
2016

You want your china, your silverware, and your set of TUMI luggage.

I registered for TUMI when I got married. And to me, it represented a rite of passage into adult luggage. As my style has changed, so has my TUMI. I've since graduated to the aluminum luggage. When I first got the new set of aluminum pieces everyone said, "Be careful with those." But I want them to scratch. They look cooler when they're scratched. I can also identify my bags by the dings and dents—but they're never really hard to find because they're the nicest bags on the carousel. You see a lot of the plastic cases but not too many of the aluminum.

I always wear the same charm necklace when I travel. It's either on me or in my bag. Recently I went to Tokyo with my 12-year-old daughter, and we brought four pieces of luggage. We planned on doing some shopping, so three were full and one was empty. We almost filled that one up, too! My husband, Kevin, has the golf bag, and my son uses the charging pad. My kids have started asking for their own TUMI luggage, and I tell them "OK, you're old enough now. You can have your own pieces."

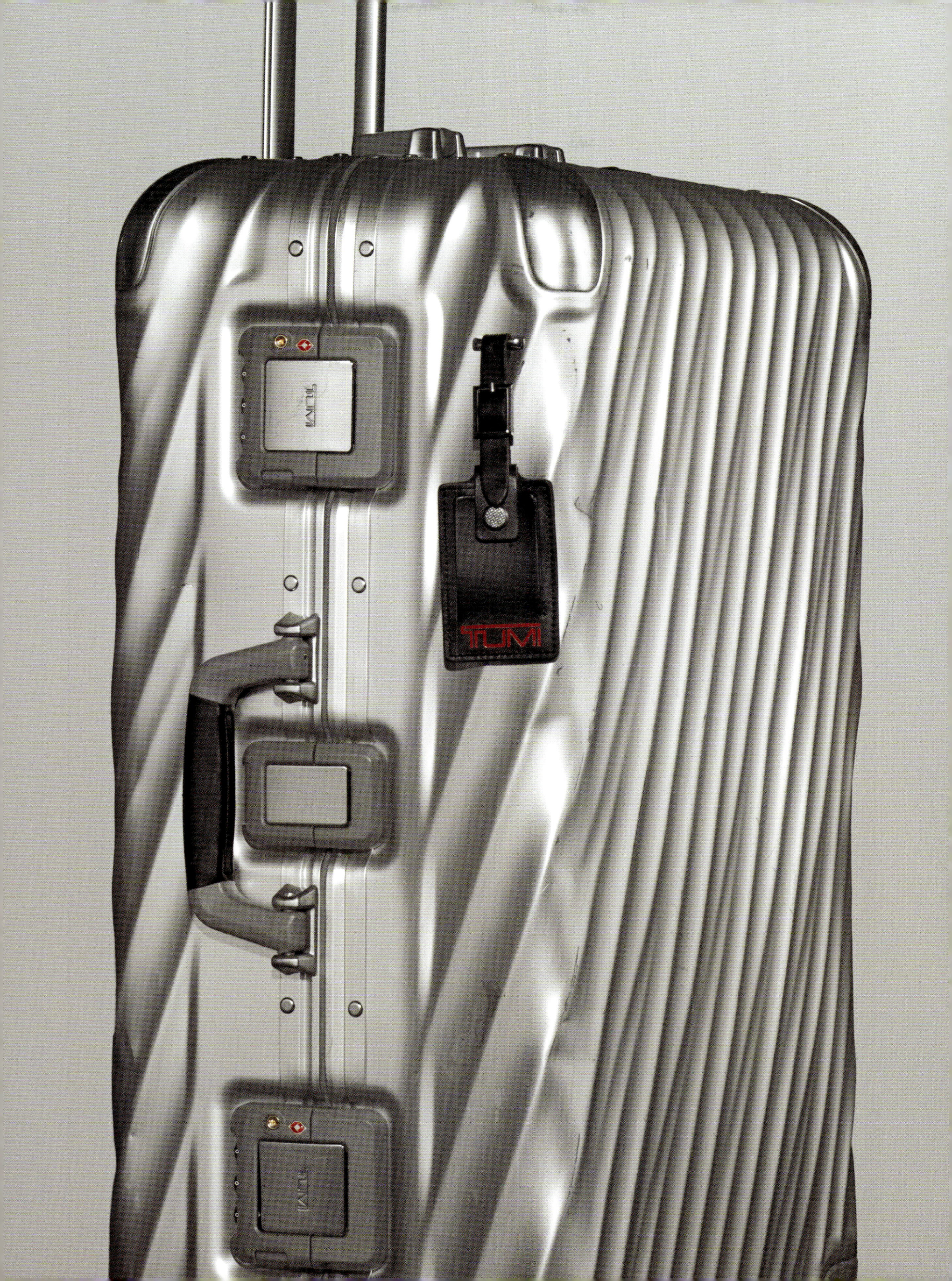
TUMI

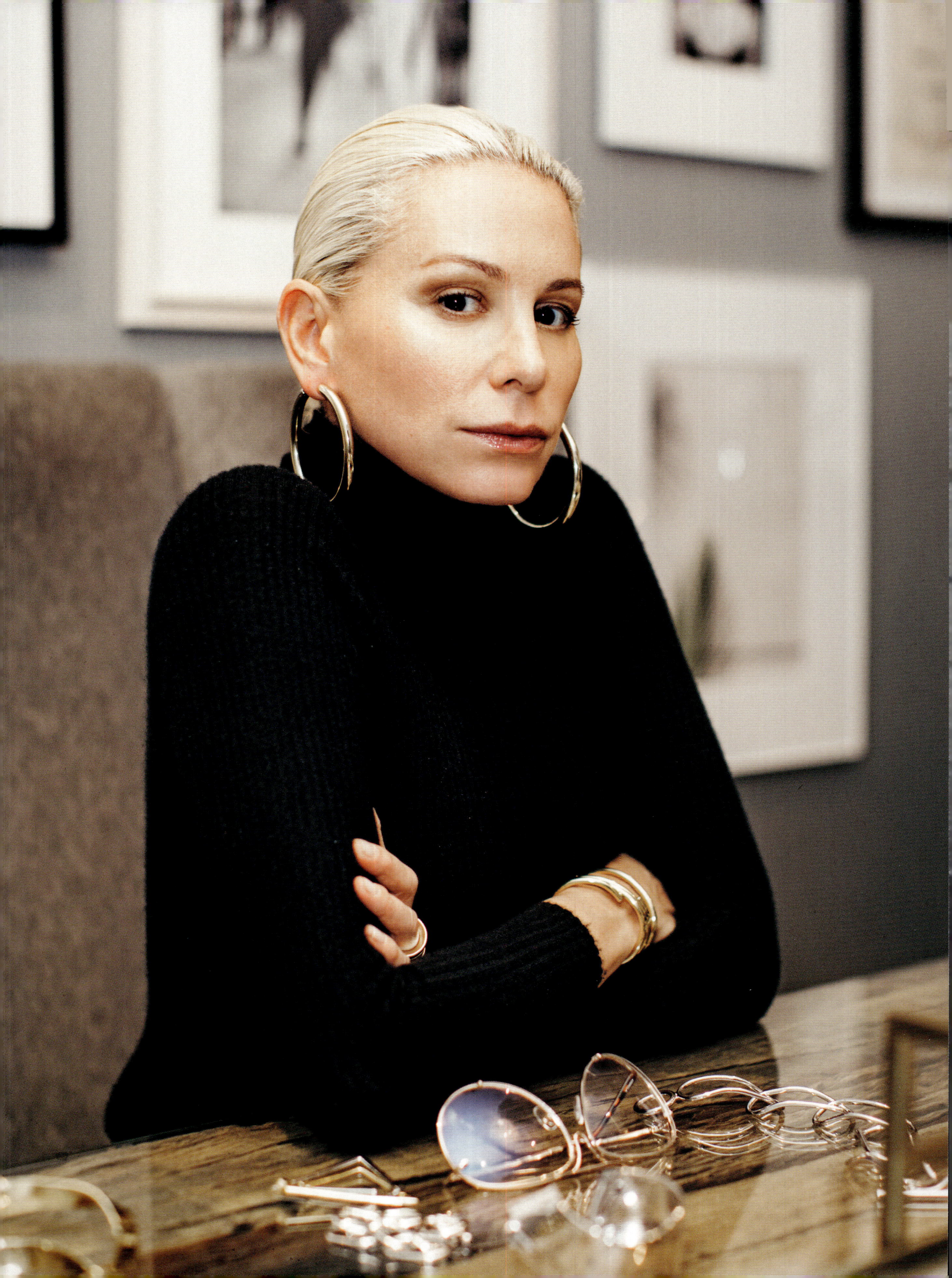

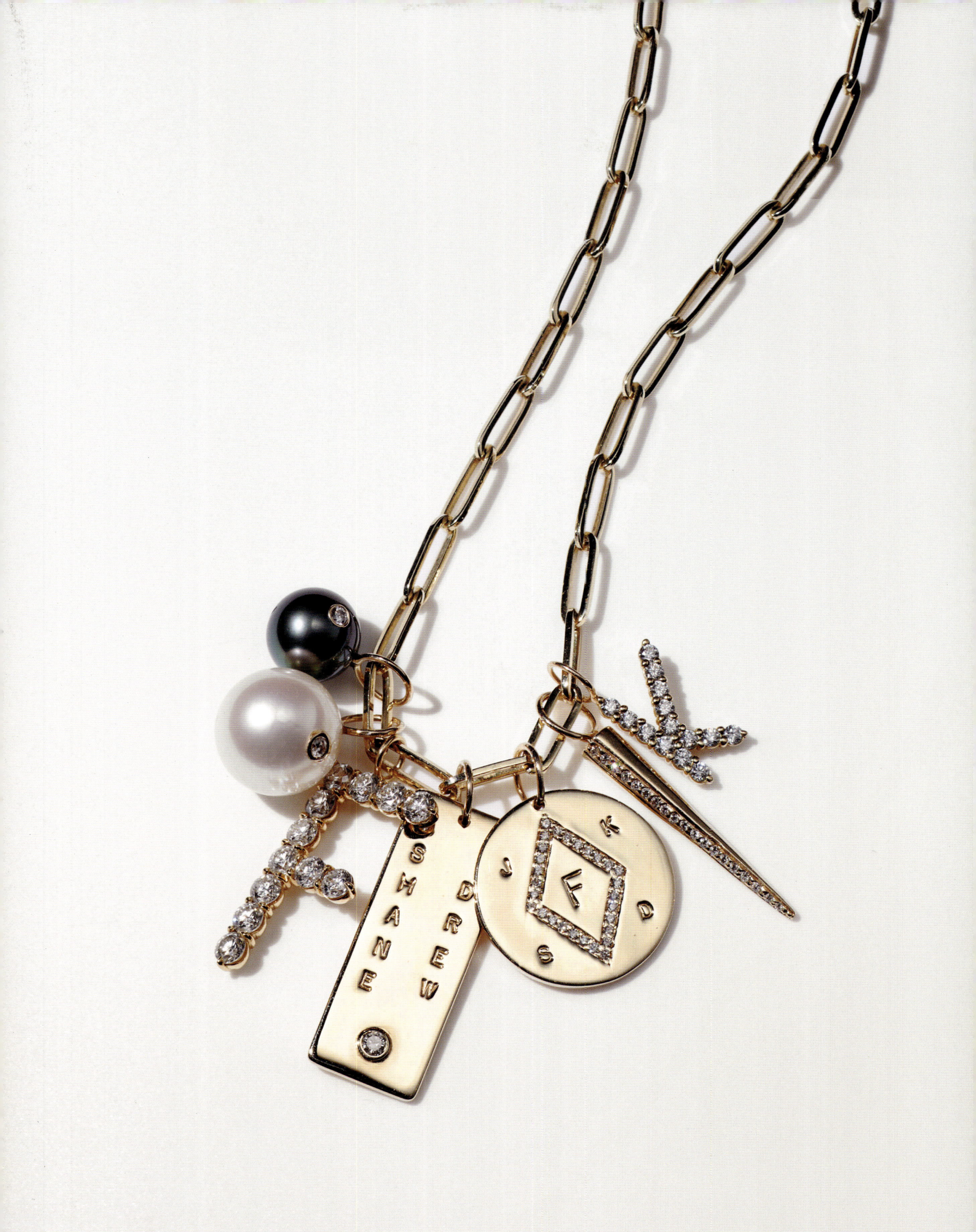
SHANE
DREW
J
K
D
S
F

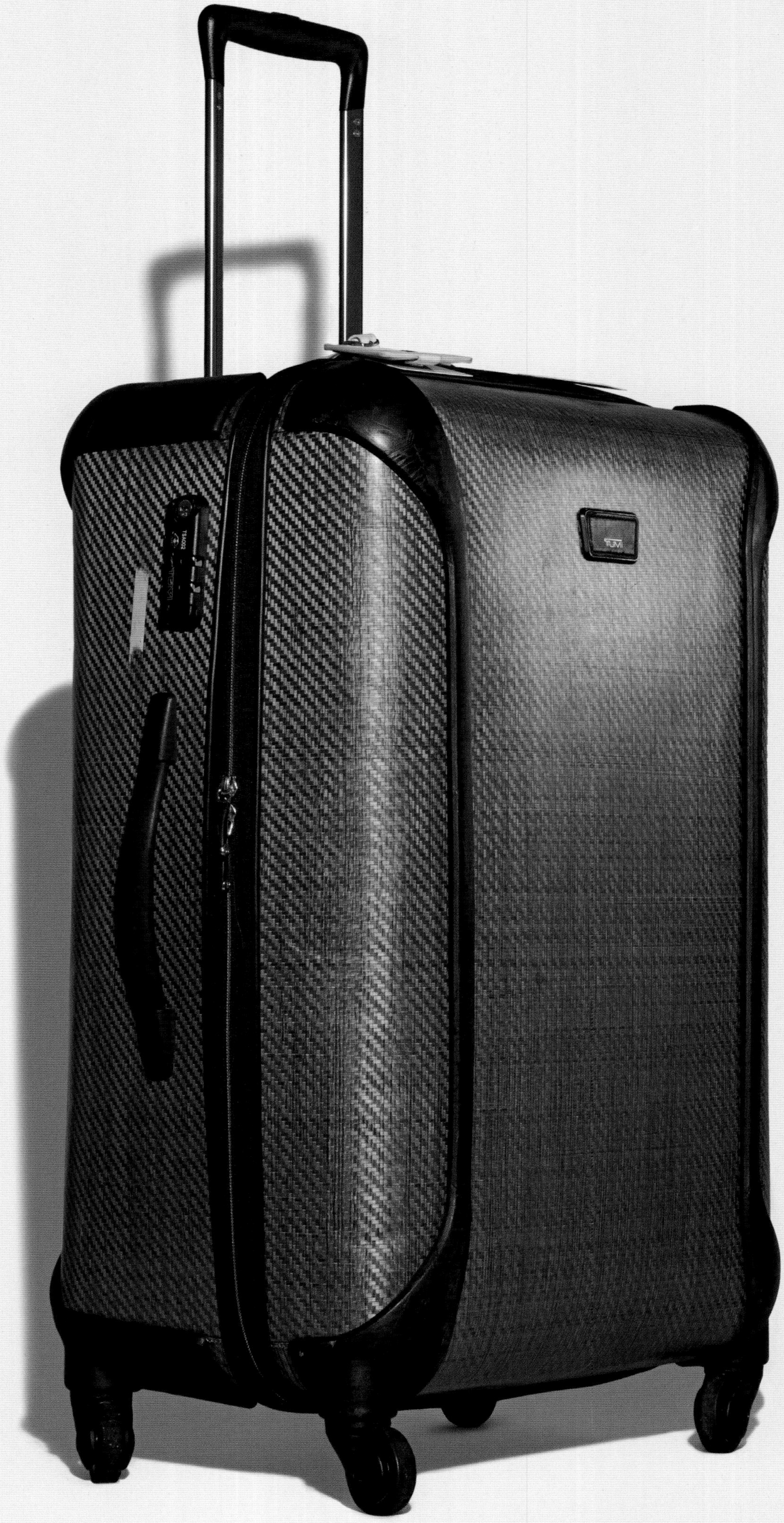

Collection

Tegra-Lite

Style

Continental
Extended Trip
Packing Case

Collection
Launched

2012

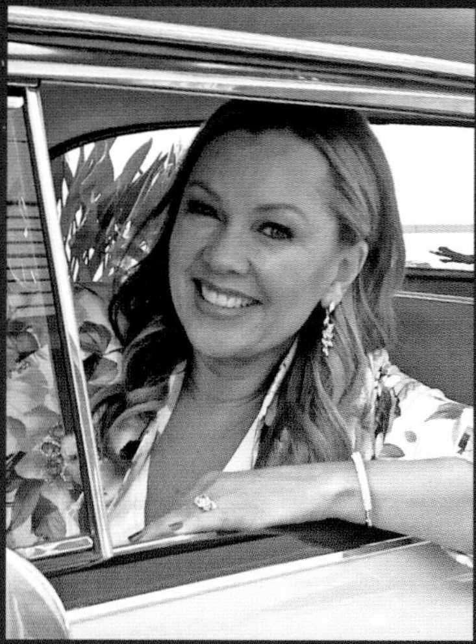

They're always with me and they always look great.

Vanessa Williams

Actress and Singer

I bring my TUMI luggage on every international flight. They've had a few dents—but I've been able to knock them out. I've been on the road for twenty years, and these have been all over the world with me. The smaller one holds my hair care products and makeup—I do all my own makeup when I perform and I do at least four concerts a month. The bigger one is for gowns, clothing, and shoes. It's my road gear for sure. They're sleek and elegant, but at the same time pragmatic and sturdy. When I leave for Singapore tomorrow, I'll definitely have my two TUMI bags with me.

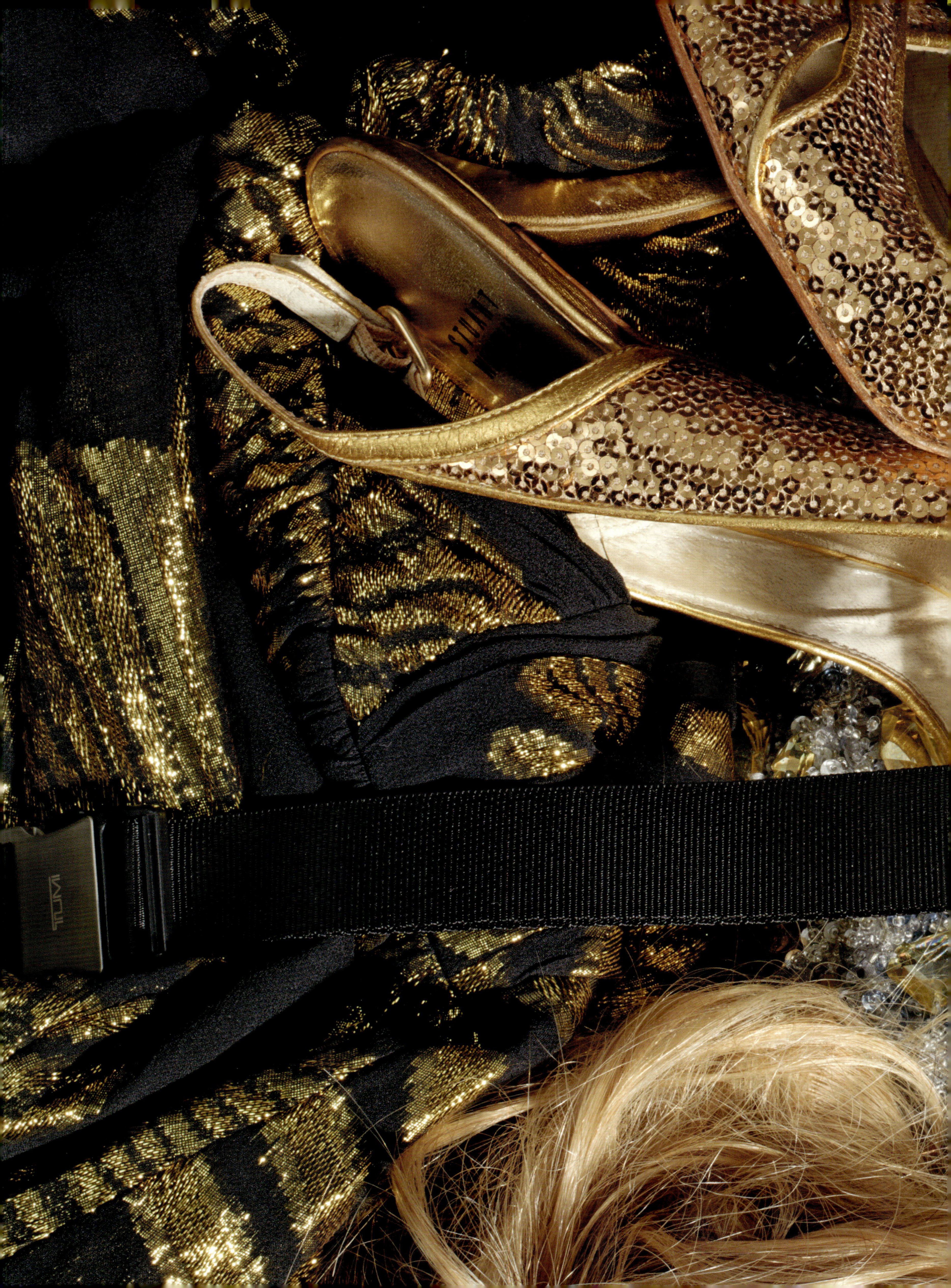

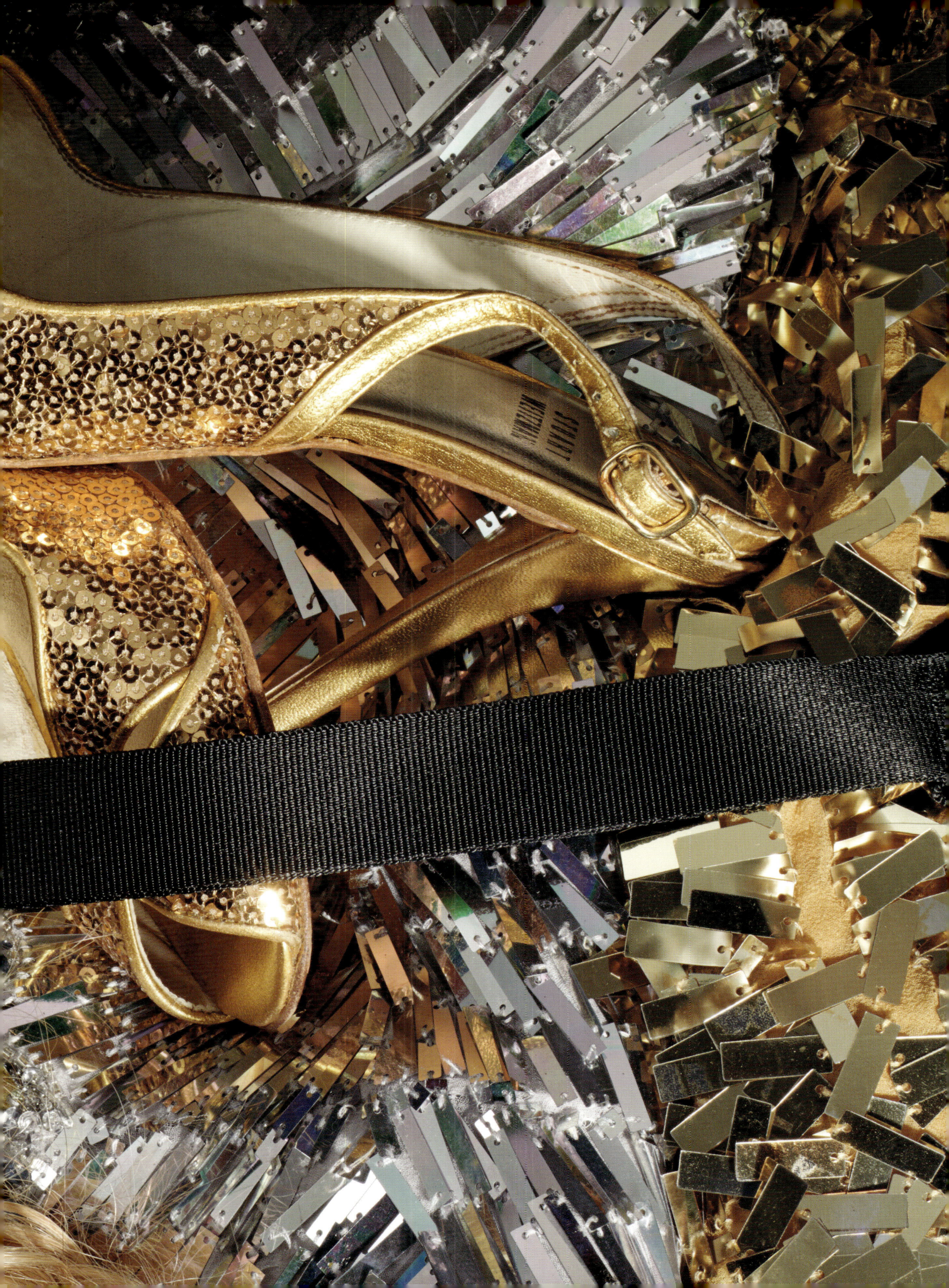

From the time that I pack to the moment I arrive at my destination, it has to be as seamless as possible.

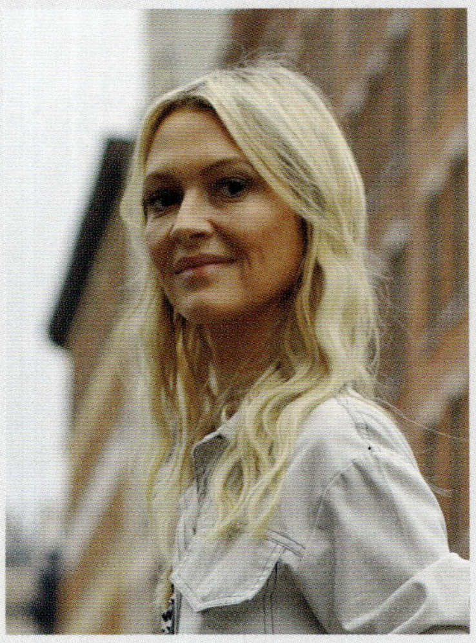

Zanna Roberts Rassi

Fashion and Beauty Journalist and Co-Founder of Milk Makeup

My first TUMI piece was a gift: a roller suitcase that was definitely a step-up from the basic luggage I had been using. I was traveling more for work, and wanted something that was understated, yet still had recognition. It was my companion for eight years. Then I got the backpack. When I had my children—twins, now four-and-a-half—I realized I could fit my stuff and theirs inside the backpack and still look chic. I used it as an alternative to a diaper bag. Packing for me is very formulaic: I bring one dress that I could wear for day or night; two pairs of pants; three tops; four accessories or pairs of shoes; five pairs of underwear; and six socks. From the time that I pack to the moment I arrive at my destination, it has to be as seamless as possible—especially when I'm traveling with the kids. For that to happen, you need extremely organized luggage, pieces that will not break. One hiccup can completely derail the experience. My TUMI helps me be a more organized packer and really a more organized person at the end of the day.

Collection
Alpha Legacy

Style
International Carry-On

Collection Launched
2008

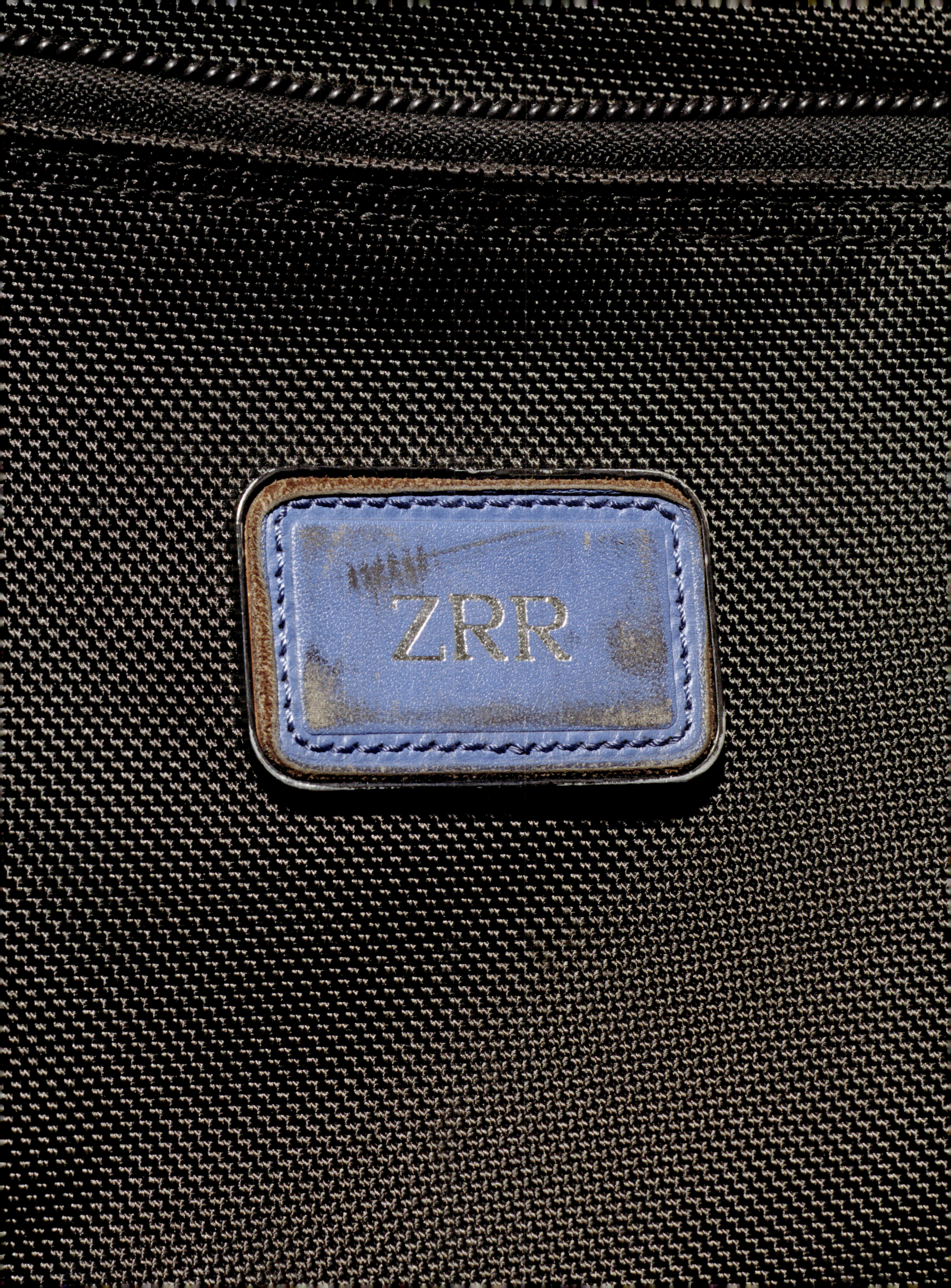
ZRR

TUMI helps me do whatever it takes to get to where I want to be.

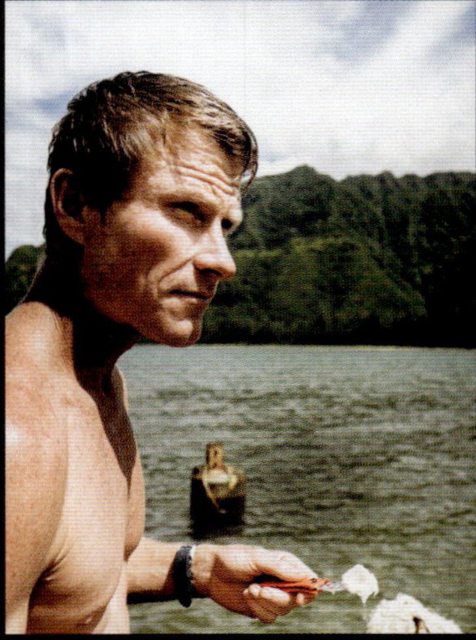

Mark Healey

Big Wave Surfer, Ocean Conservationist, and Overall Waterman

Collection
TUMI x Healey Water Ops

Style
Survival Duffel

Collection Launched
2019

I would pretty much characterize myself as a professional traveler—and yet, my style of traveling is so asymmetrical. Within one travel experience, I can go from sleeping on the dirty floor of a tiny airport to having a meeting in a five-star hotel, and, until I found TUMI, there weren't really bags that fulfilled both sides of my travel experience. TUMI products excel in all of the areas I need when I'm traveling—durability, functionality, and especially an eye towards design.

Working on our collaboration, The Survival Duffel, was an incredible experience. Being part of the design process with Victor and the team, who are some of the highest level designers I've ever worked with was so much fun. They were truly open to what I had to say, even though I have no official background in design, and ran with it. We wanted to create this piece that was functional and raw round the edges but could also fit in easily in first class, and I think we did just that. Pulling out your ID, wallet, cash or laptop is so easy because you can just swing the bag around your body, especially in places where you don't want to put your bag on the ground. And making it from recycled materials was important to me because it's a step in the direction of proving that a recycled product can look and perform just as well as one that isn't. One of my favorite stories, which exemplifies my relationship with my TUMI bags, was when I had to go to the Big Island for a meeting by way of a small boat with waves crashing and water spraying. I tied my TUMI Survival Duffel about 6 feet up the mast, but still it got totally soaked. When I took my bag down, I thought for sure everything would be wet, but to my surprise everything in my bag was dry! That's what I love so much about my TUMI bags, they fit my odd style and stand up to the harshest of environments. They look and perform the way they need to whether it's traveling to an island in the middle of nowhere or going to an upscale event—that is what is most special about TUMI's products to me.

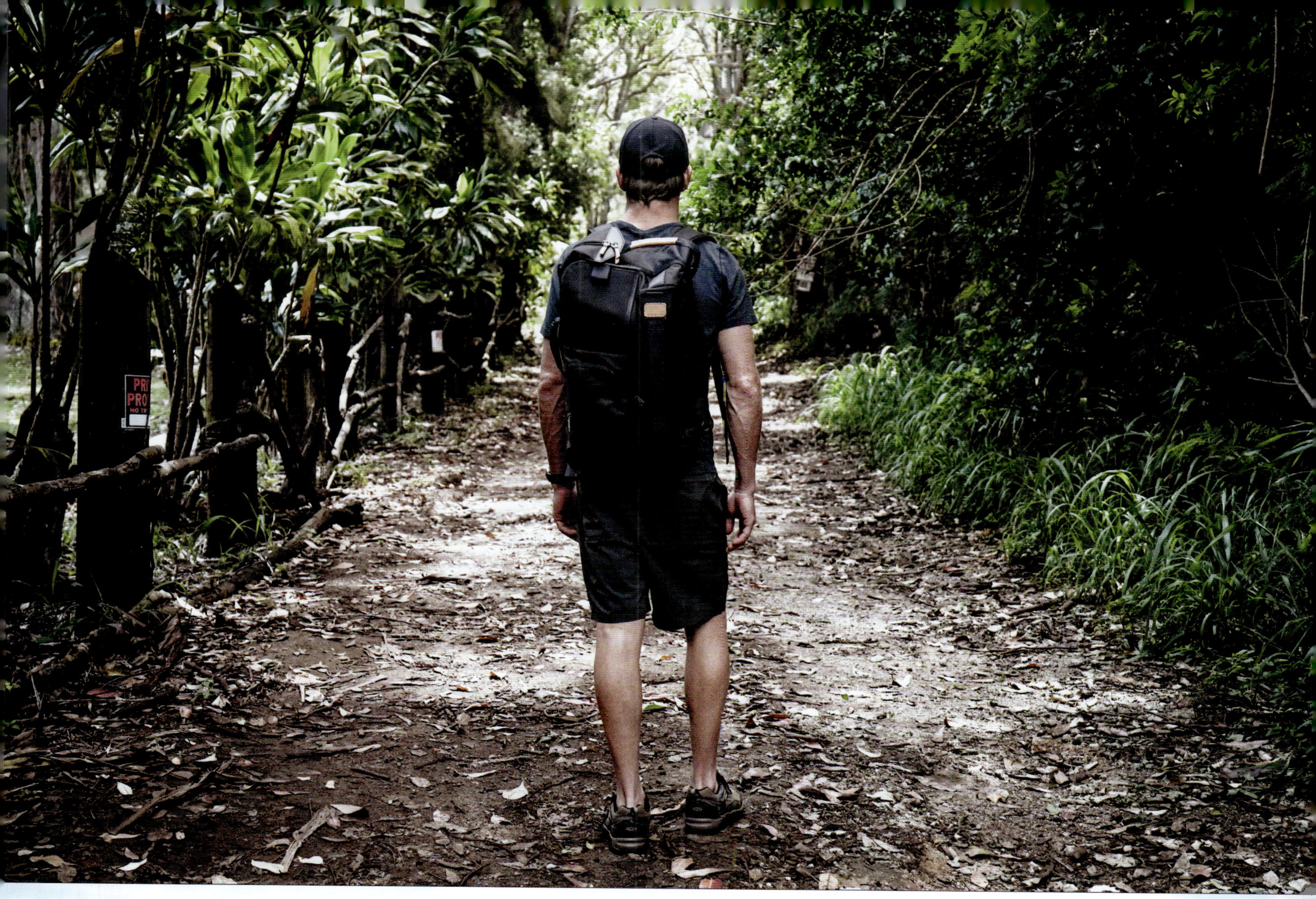

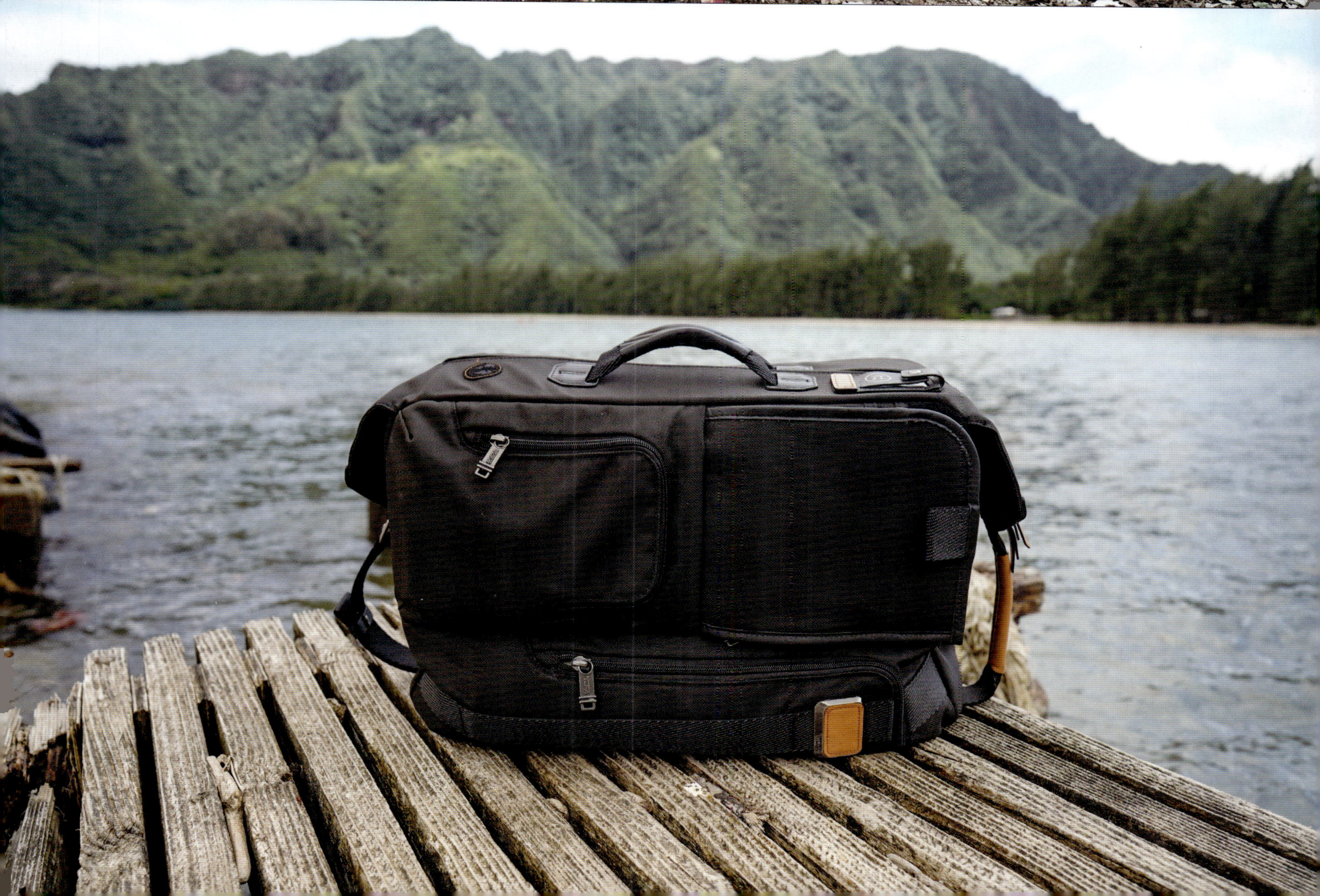

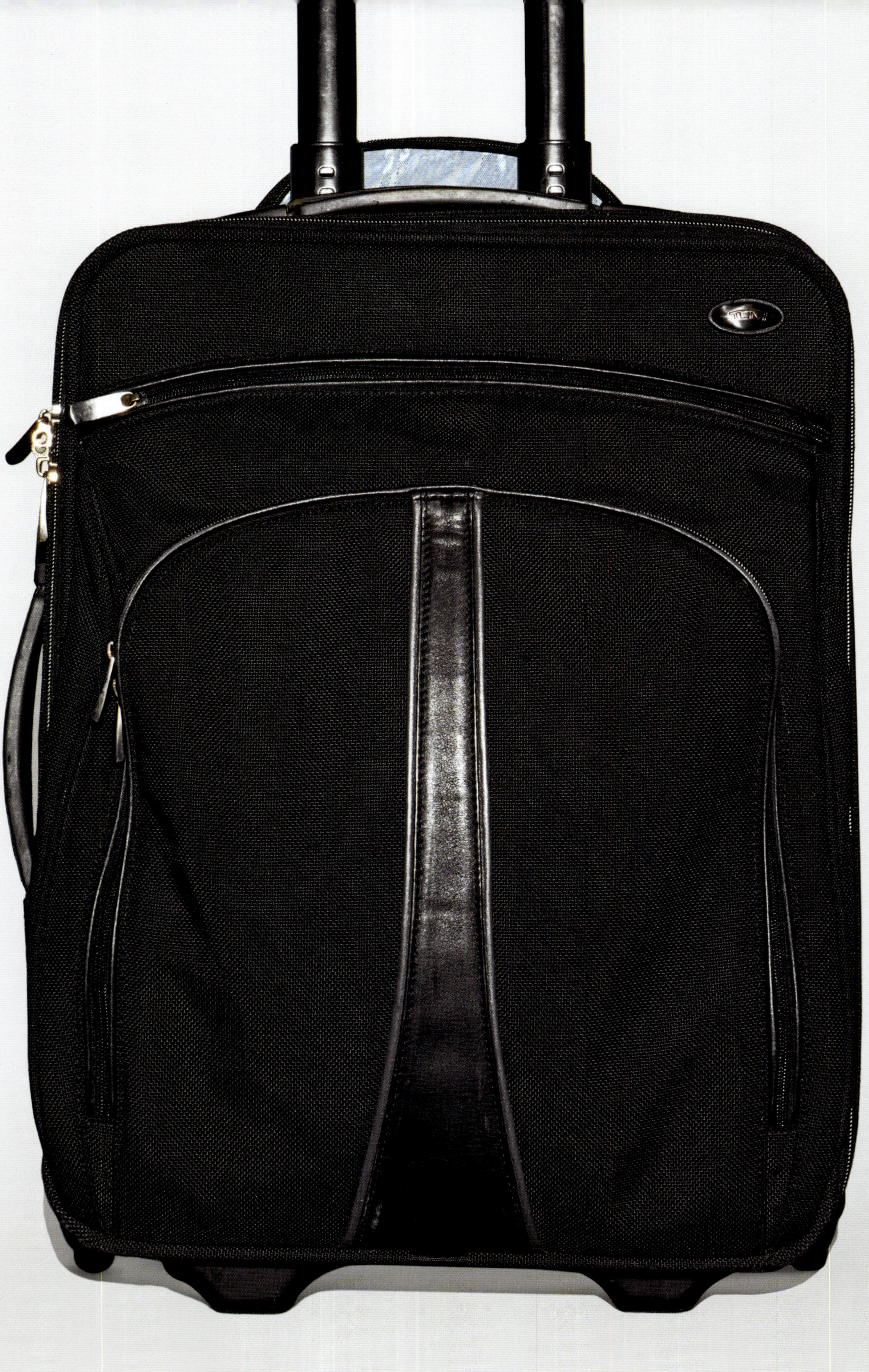

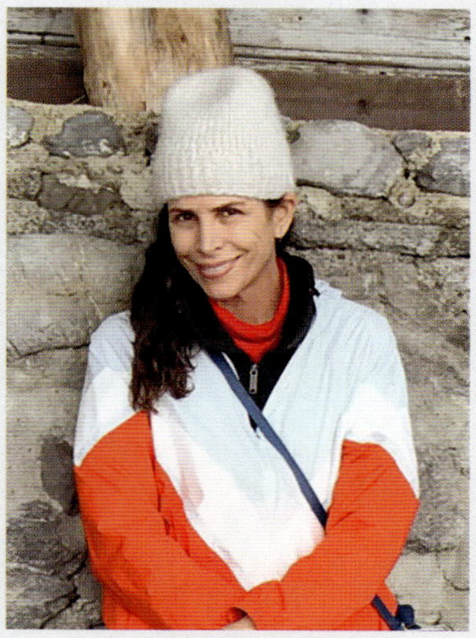

Collection
Elements

Style
Expandable
Wheeled
Packing Case

Collection
Launched
2006

I don't just use it when I travel, I carry it every single day.

Cristina Ehrlich

Celebrity Stylist

I've been a celebrity stylist for 20 years—and bicoastal for 15 of those years. Anyone who knows me would say that through all of the apartments I've lived in, and through all of my travels to festivals and fashion weeks in different cities, my most consistent relationship has been with my luggage.

Clothes are a very important part of my life. My mother is from Bologna, and when it comes to the traditional, classic Italian woman, fashion, dressing, travel, and packing, you have to do it in a specific way. My mom taught me that when you travel, you cover your clothes in plastic and hang them on a paper hanger, then lay them flat in your suitcase. Pretend it's a lasagna. You layer, so that when you get to your location, you may just need a gentle steam.

I've tried luggage from just about every brand and the suitcase I covet the most has always been my TUMI. My "Portable Office" is my fashion wubbie. I take it from my home to my office, which is a six-minute walk. There's a compartment for my laptop and my daily planner; a section for toiletries; and a zip pencil case where I store my chargers and cords. I pull this bag all over the place, then she gets unpacked every single night. She may not look like a fancy pants celebrity stylist's suitcase. But in a way, I like that.

We probably had about 20 seconds warning prior to landing, and we didn't know that we'd be landing in the water.

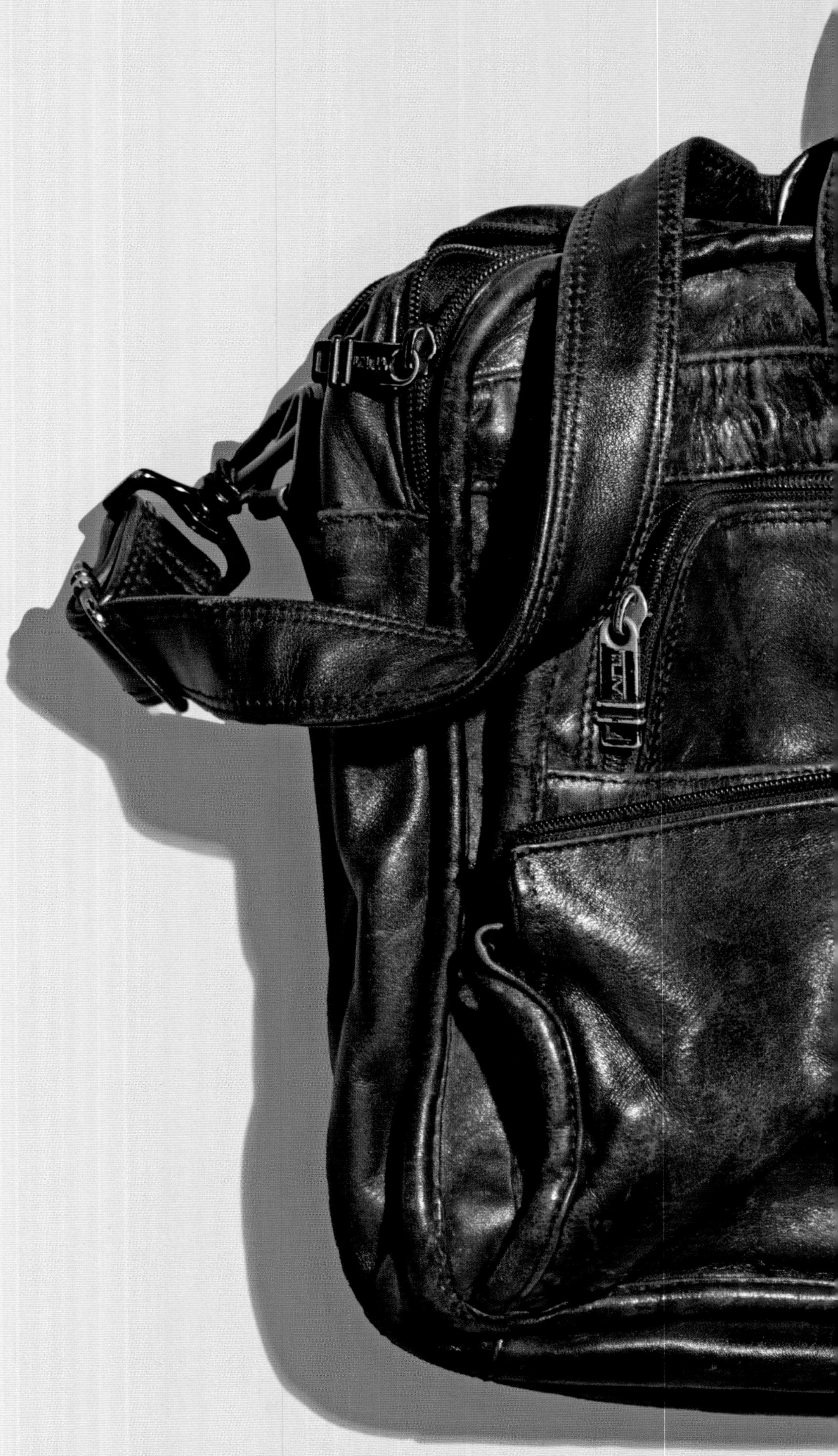

Kevin Anton

Passenger on US Airways Flight 1549, aka "Miracle on the Hudson"

I thought everything was looking pretty good until Sully said "Brace for impact," then I thought, oh shit this isn't going to be good. Once we hit the water, it rushed in immediately. It was waist-deep in about 20 seconds. All you're thinking is how do I get my butt out of here so I don't drown in this tube.

The weekend after the flight, I went to the local mall and picked up a new TUMI briefcase and roll on, and I started using those. But US Airways ultimately got everything off the plane and sent our luggage to a firm that specializes in salvage and recovery. They cleaned it all. They went through my binders and dried every page. Everything was wrapped in dryer sheets when they shipped it back to me a few months later. As soon as my old TUMI luggage came back, the new ones went up on a shelf. I prefer to use the old ones. They're resilient and durable. Kind of like an old friend.

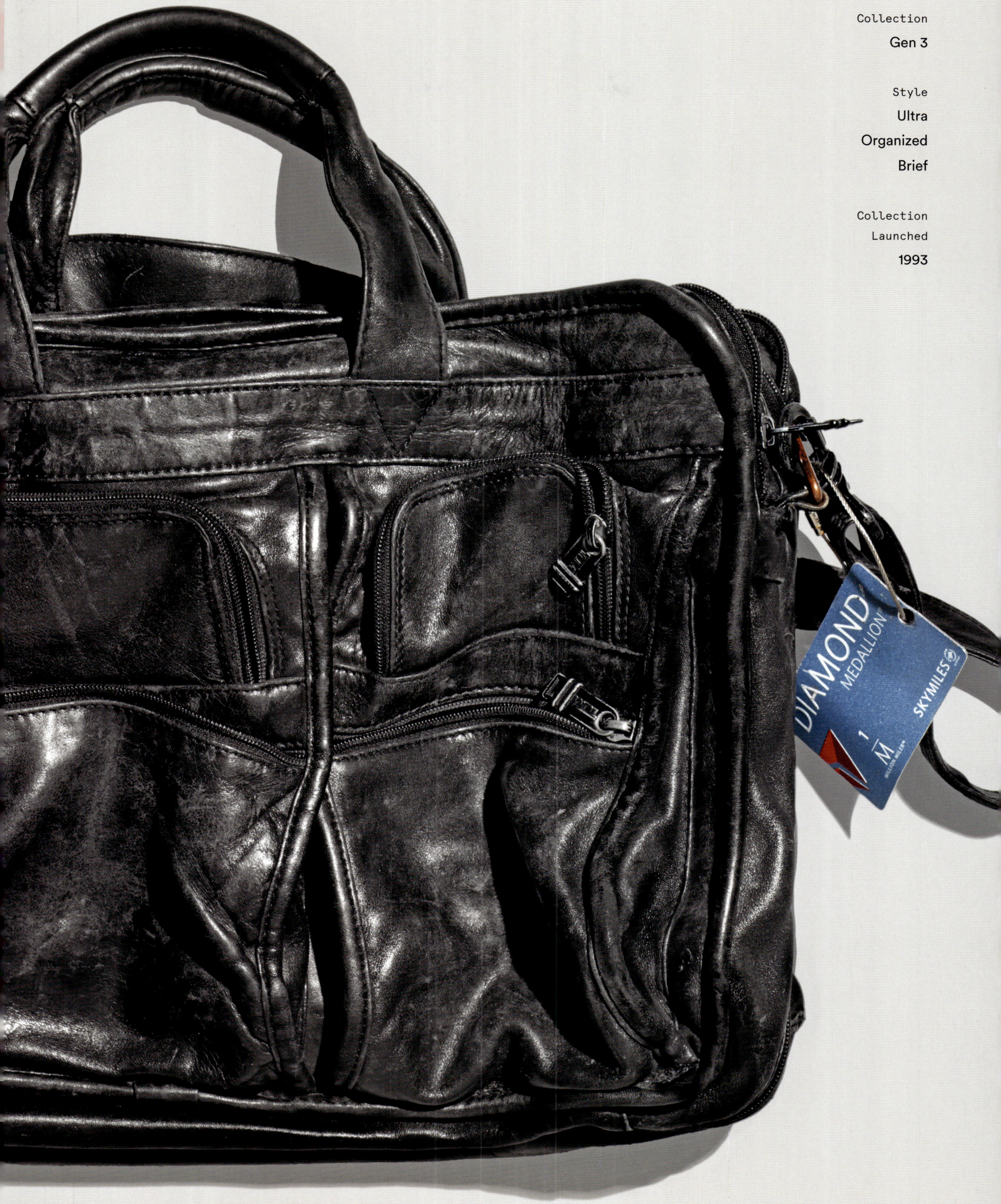

Collection
Gen 3

Style
Ultra
Organized
Brief

Collection
Launched
1993

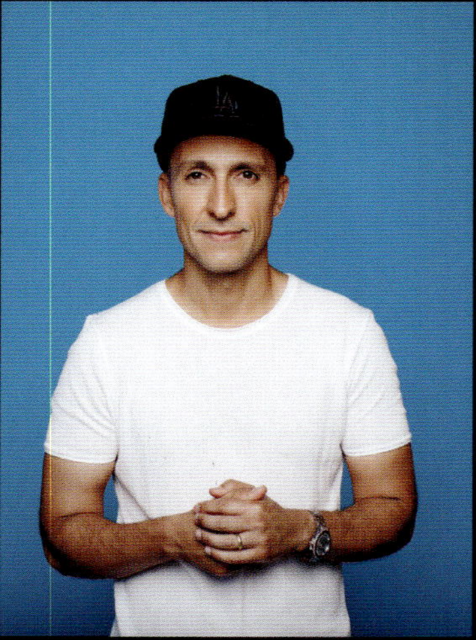

Vice

Los Angeles-based DJ

Collection

TUMI x DJ Vice

Style

Brief Pack

Collection Launched

2013

This bag has traveled with me more times than my wife and kids.

I came across TUMI when I hit status and started clearing upgrades. All of a sudden I was hanging out with elite and business travelers and I began to see TUMI luggage all over the lounge. Everyone's bag was in such good shape. The design was sleek and to the point. Everyone had on sport coats. I didn't. Everyone had on dress shoes. I didn't. I was in running shoes. But now I have a TUMI bag just like everyone else. And I know I won't look awkward rolling into a nightclub for a gig with my carry-on luggage and backpack. It plays to my style. I do more than 250 shows a year. This bag has traveled with me more than my wife and kids. It's taken me around the world and back many times.

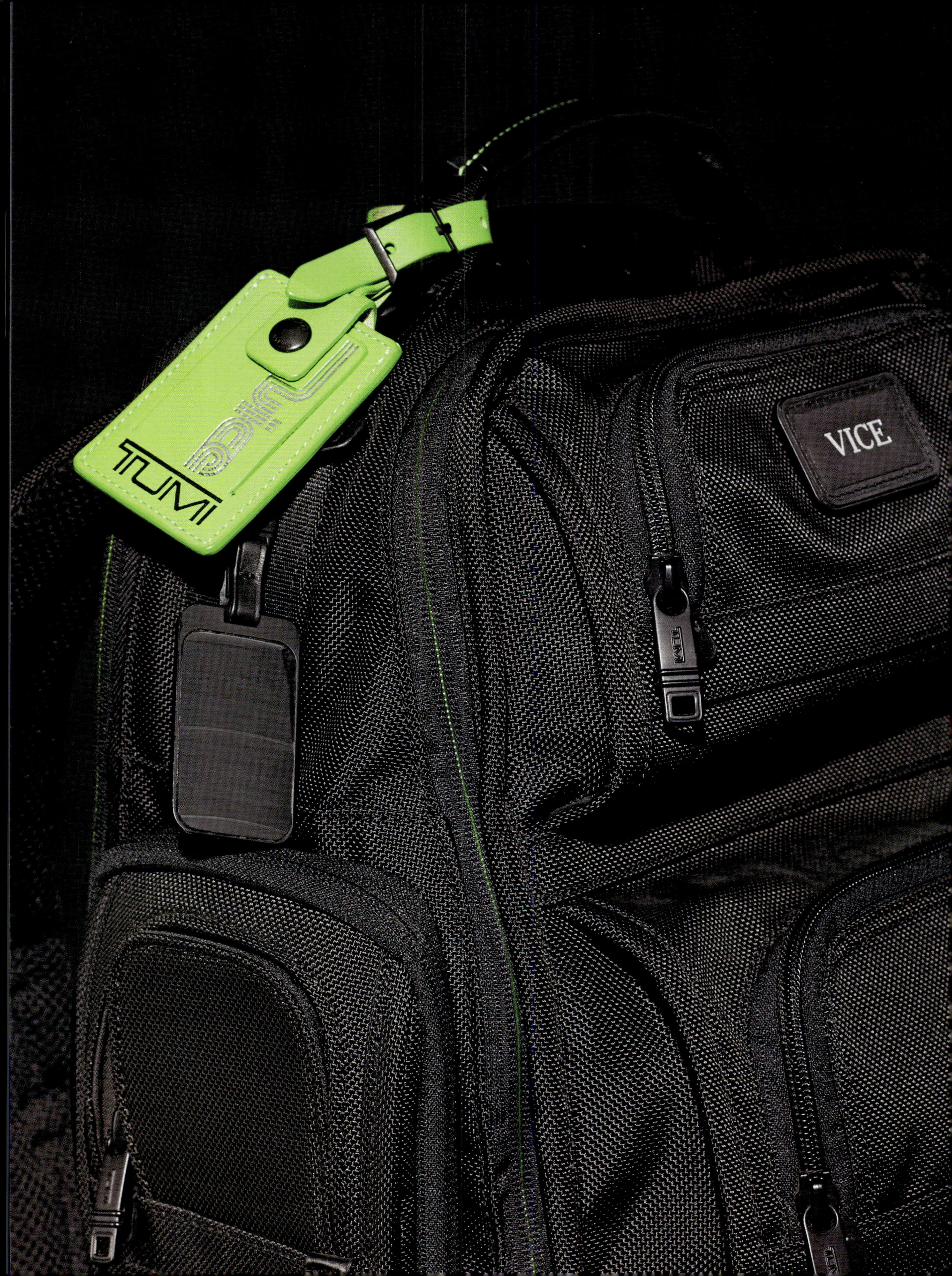
TUMI
VICE

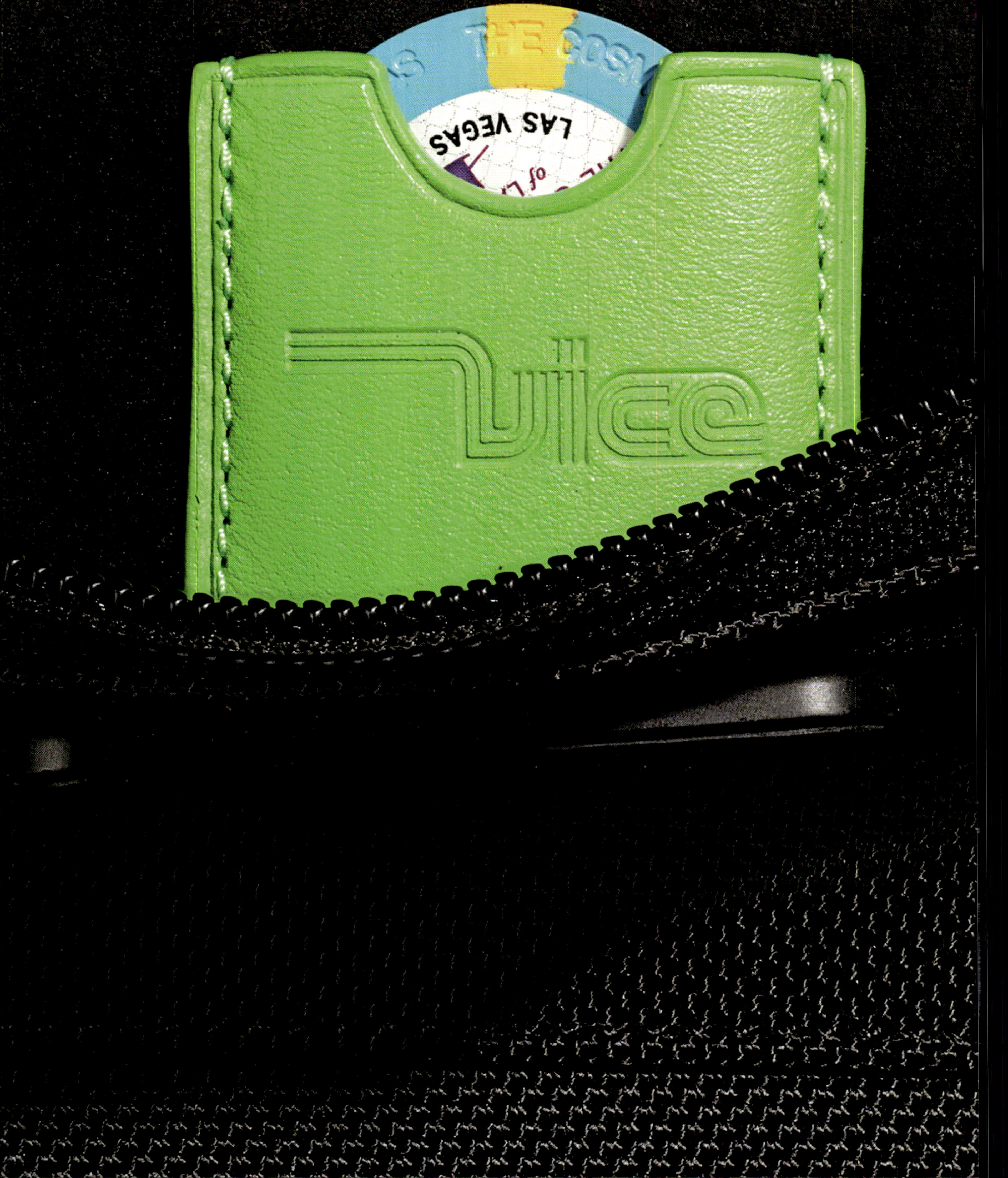
LAS VEGAS

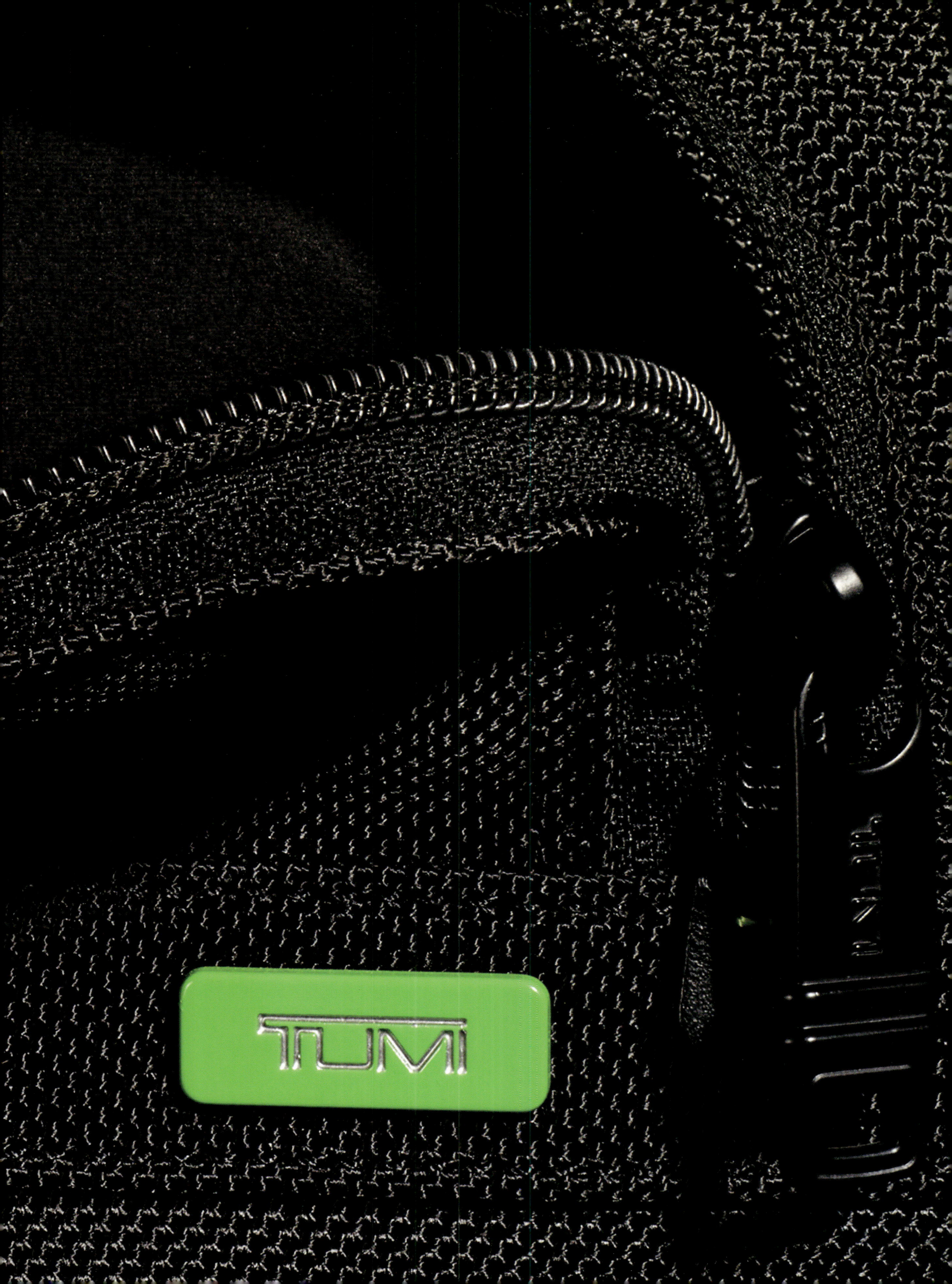
TUMI

BACK

Collection

Alpha Legacy

Style

Extended Trip
4 Wheeled
Garment Bag

Collection
Launched

2008

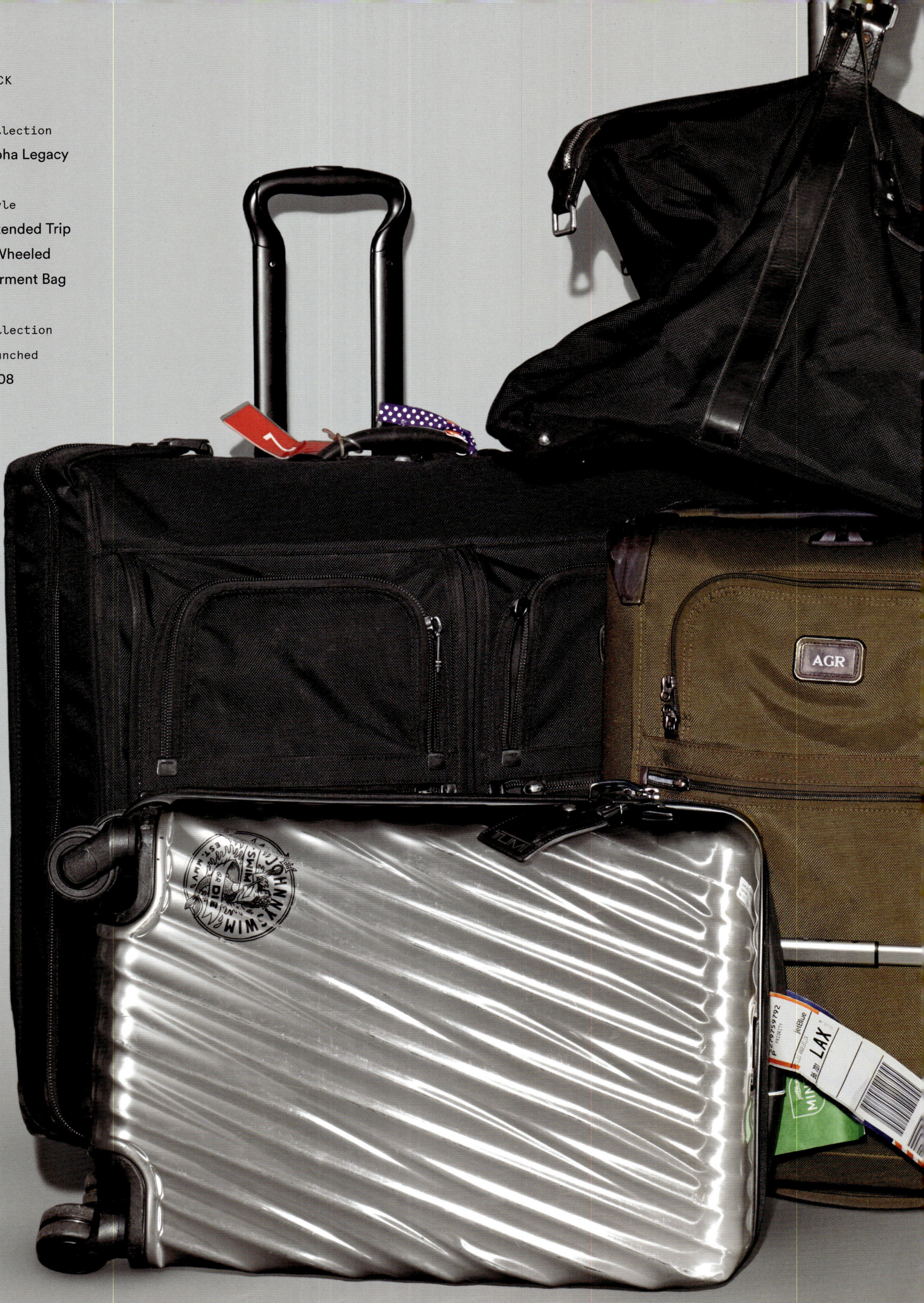

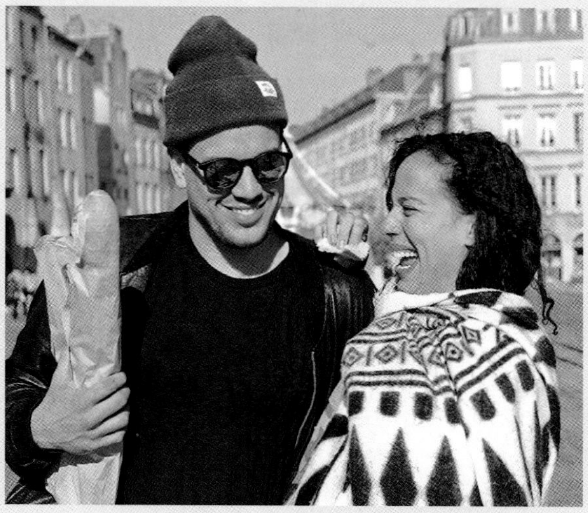

Amanda Sudano Ramirez & Abner Ramirez

Duo behind the band JOHNNYSWIM

Amanda: My parents were musicians too—my mom was a singer named Donna Summer—so they toured all the time and they're the ones who introduced us to TUMI luggage. To this day, if I'm looking for certain piece, I'll go to my dad's house and find it there. It's like a store.

Abner: I didn't travel all that much when I was younger—we were a poor family from Jacksonville, Florida. I was swept up by all of these feelings when I walked into Amanda's parents' room of TUMI. To me, those bags represented freedom—freedom to travel, to see new things, meet new people, and explore new cultures. It meant living life fully.

Amanda: There are three pieces in particular that we have a connection with. One is my green bag. I've actually been meaning to take it in for about two-and-a-half years now but I don't want to be without it for any period of time. I love it. It knows me and I know it.

Abner: I have a duffel that Amanda's parents gave me. Actually Amanda's mom, for my 25th birthday, gave me the fanciest gift I'd ever received in my life until that point. She got me a TUMI weekender bag. It was my first real piece of luggage.

Amanda: And now that our son Joaquin is four years old, he's big enough to have his own TUMI. His has a smiley face on it so he knows it's his. He pulls it all by himself.

Growing up, when we got to a certain age, we got our own TUMI bags. It was like a rite of passage.

FRONT

Collection
19 Degree PC

Style
International Carry-On

Collection Launched
2016

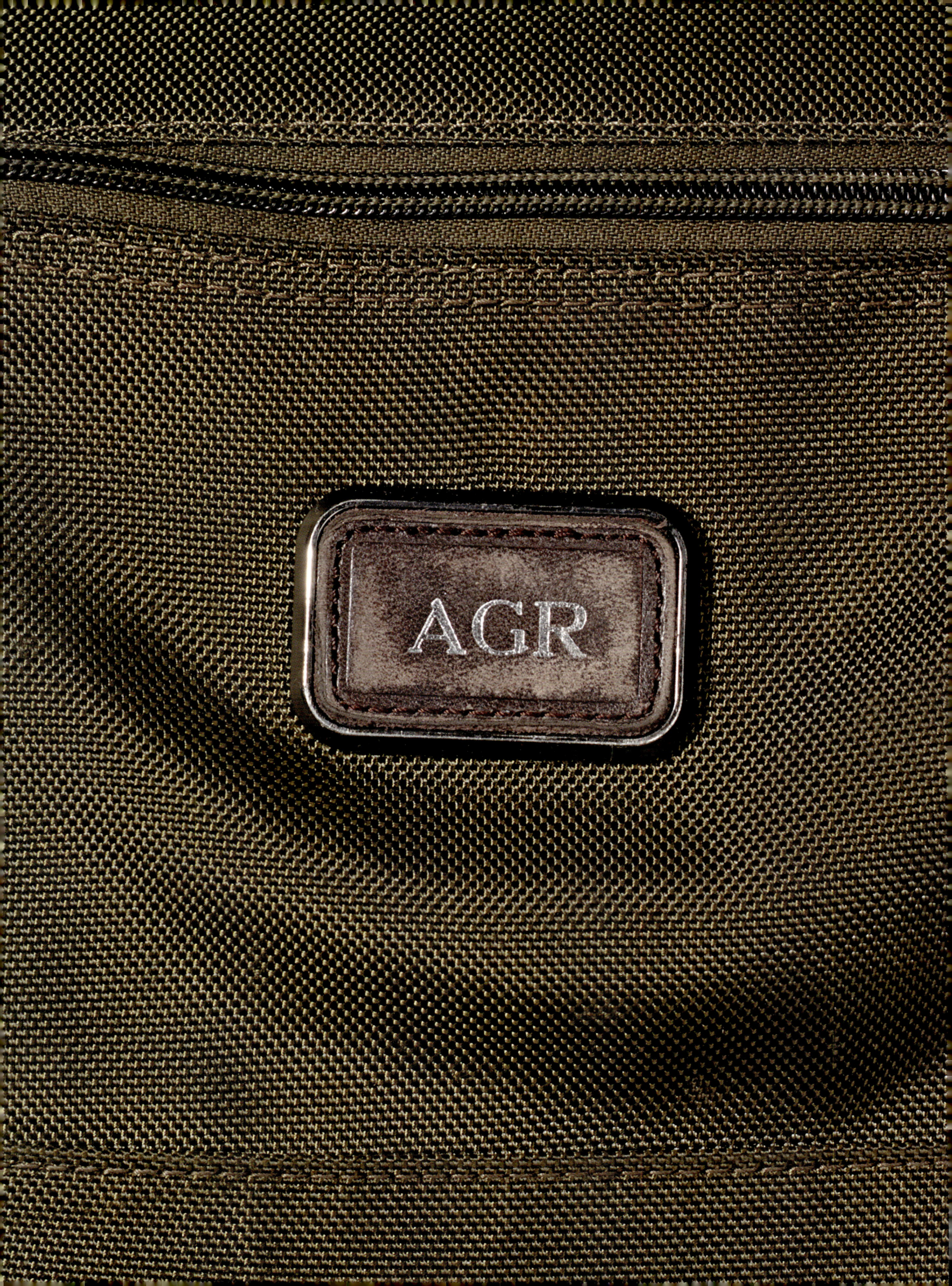
AGR

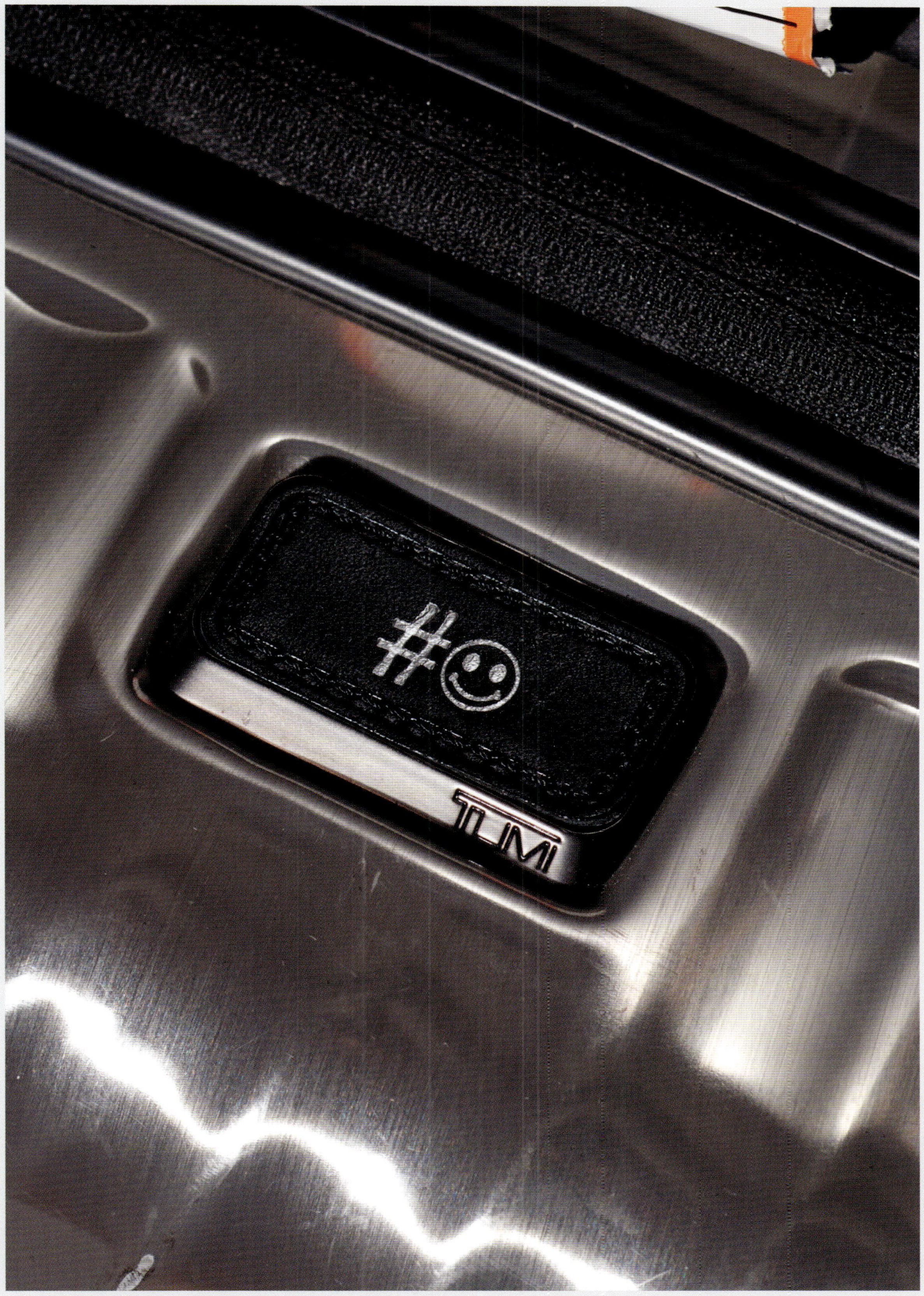
TUMI

TUMI is the one brand that feels like an extension of my home.

Collection

Harrison

Style

Webster Backpack

Collection Launched

2017

Daniel Henney

Film and Television Star

I first discovered TUMI years ago when I was living in Hong Kong. I didn't have a lot of money at the time and I was in need of new luggage. One of my friends gave me a TUMI carry-on for my 22nd birthday and I've been a huge fan ever since. I've been traveling for so many years now—and have used so many different brands—but TUMI is functional and rugged and it coincides beautifully with whatever style of clothing I'm feeling that day. It's simply the perfect bag and I do feel an emotional connection with it. It's all I have out there sometimes. When you travel as much as I do, that bag becomes very important to you. It protects very important parts of your life. No matter where you go in the world that bag has pieces of home inside.

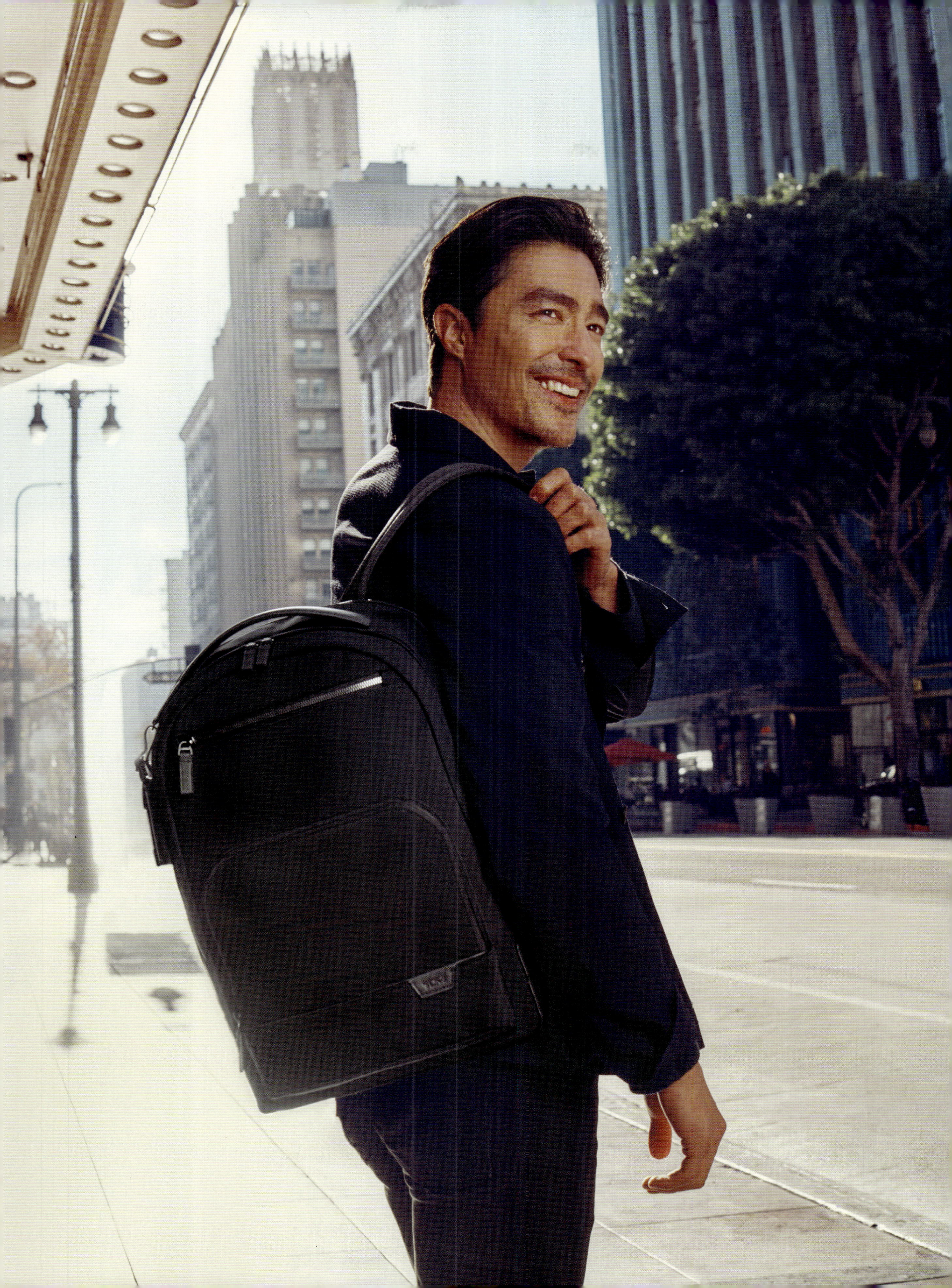

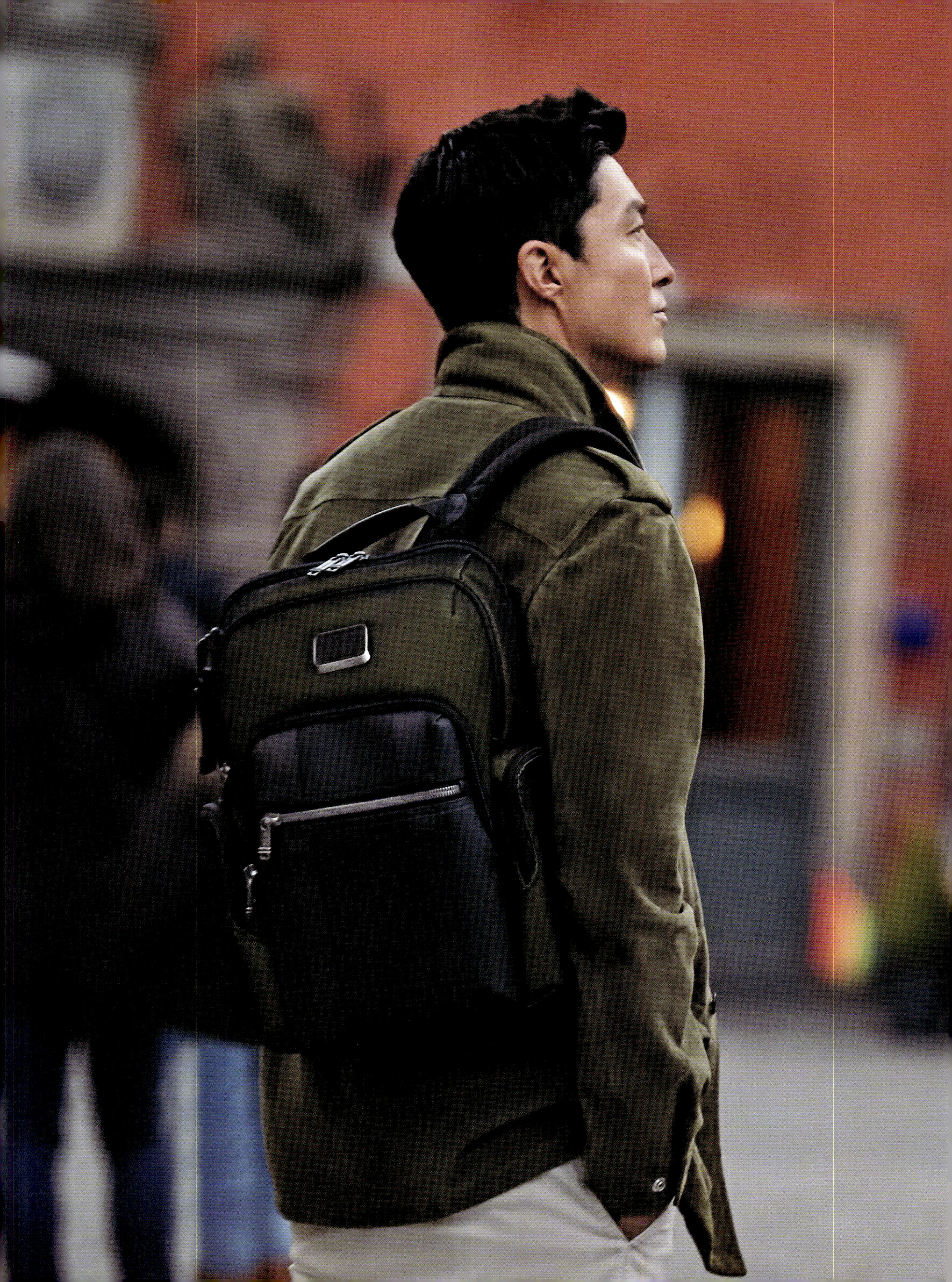

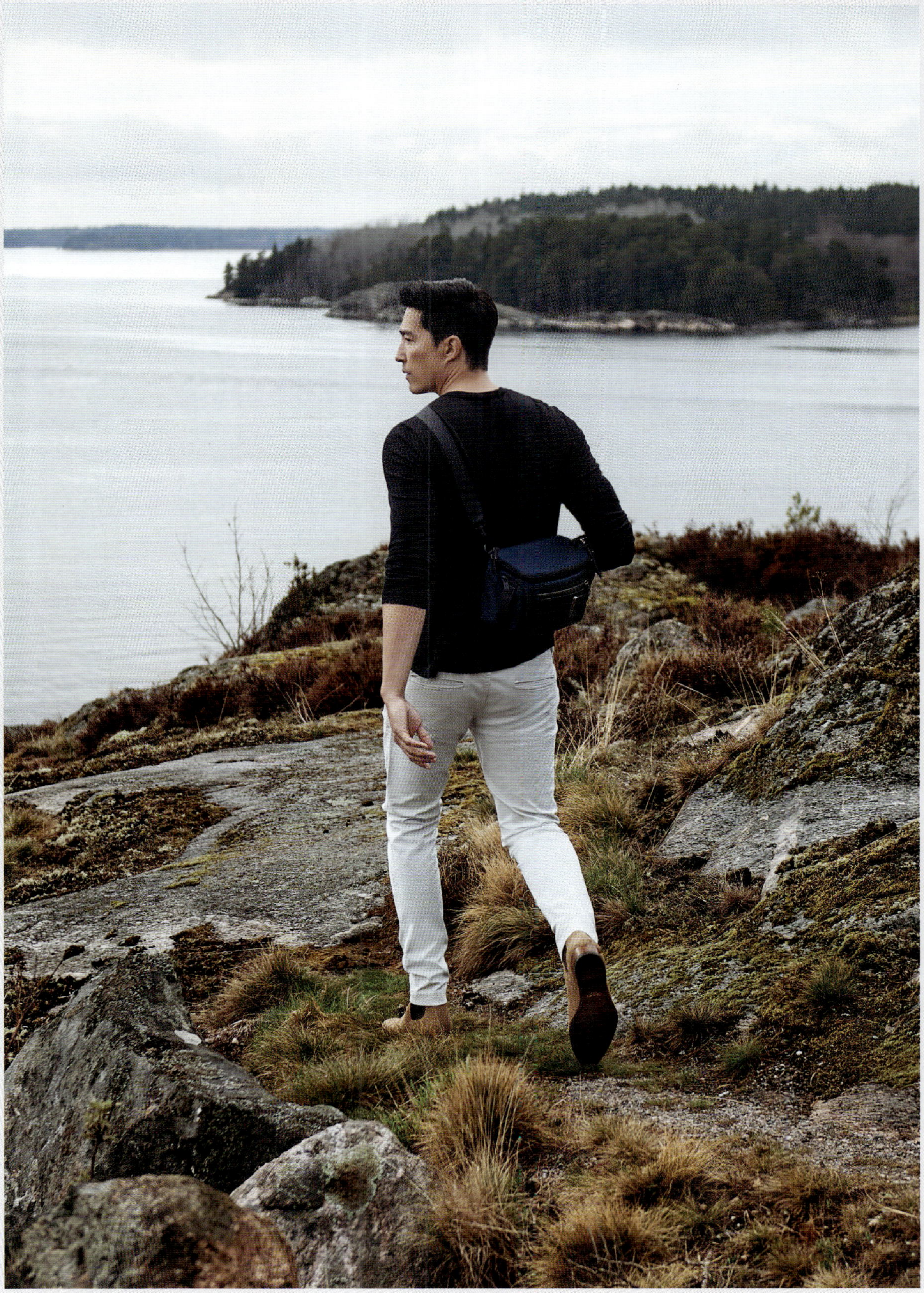

TOP

Collection

Beacon Hill

Style

Rever Brief Pack

Collection Launched

2013

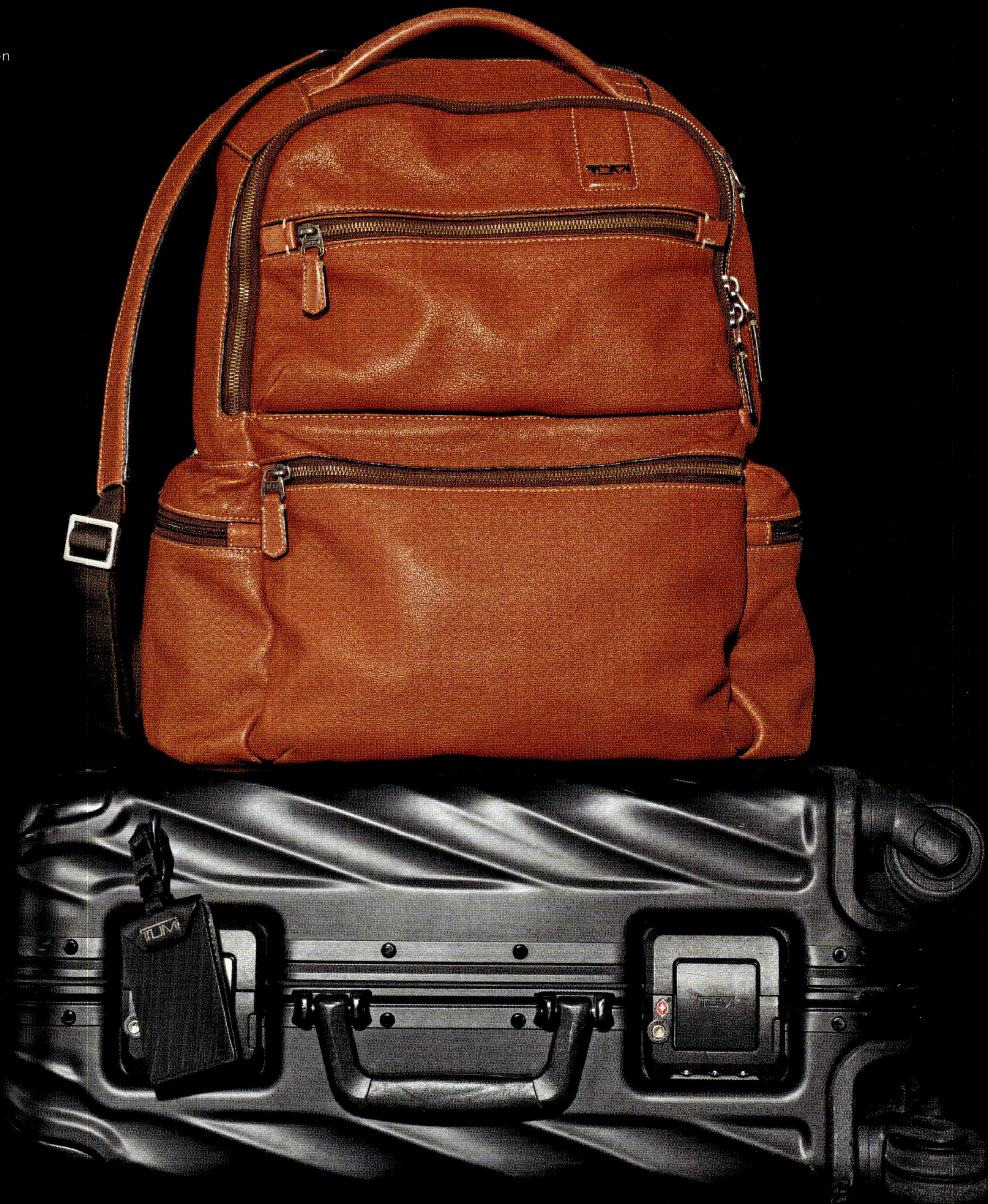

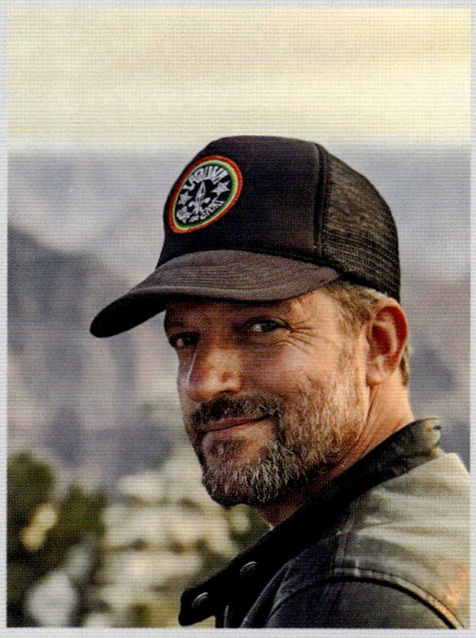

Jon Rose

Surfer, Humanitarian, and Founder of Waves For Water

Luggage can make your experience of traveling better.

There are nuances to this luggage that I appreciate. When you can literally roll your bag with your pinky finger, causing no stress to your body, that's huge. But that's something TUMI's perfected with their four-wheel system. I spend a lot of time on planes and jet lag takes a toll, so there's actual health benefits to having a TUMI bag. My body is definitely my temple, so I'm always trying to offset the negative aspects of traveling: I take supplements, stretch, maintain my workout routine. I can endure rigourous travel because I do all of those things—I'm not going to throw my shoulders and back out by lugging bags through the airport just to get where I'm going. The way I look at it, buying a TUMI backpack, like the Tahoe, is the same as buying a computer. It's a non-negotiable that I need for life. It brings me comfort and ease when I'm out there tackling the world.

BOTTOM

Collection
19 Degree Aluminum

Style
International Carry-On

Collection Launched
2016

Elizabeth Stewart

Red Carpet Stylist

As a stylist, I have incredible access to carry on bags, but I really cannot live without my little leather carry-on.

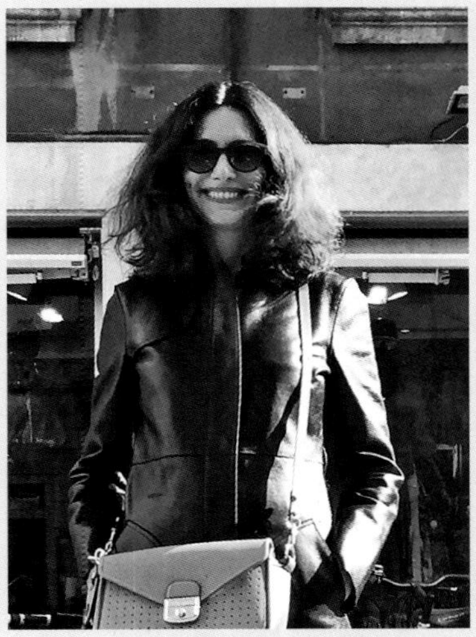

I've had it for maybe 20 years. It's slightly beaten up, but it just feels...right. It's a very simple design, but I can pile everything in it. I've even managed to pack for a four-day trip with it. I feel better when I'm traveling really light—I hate having too much stuff with me. My travel philosophy is—be zen. There are so many curve balls when you're traveling, so I always try to turn a negative into a positive: Long flights become an opportunity to catch up on books or movies that I never get to watch. A ride to the airport is always a call to a friend who I normally don't have time for. I try to use this time as down time, to be zen.

Collection
Gen 3

Style
Carry-On

Collection Launched
1993

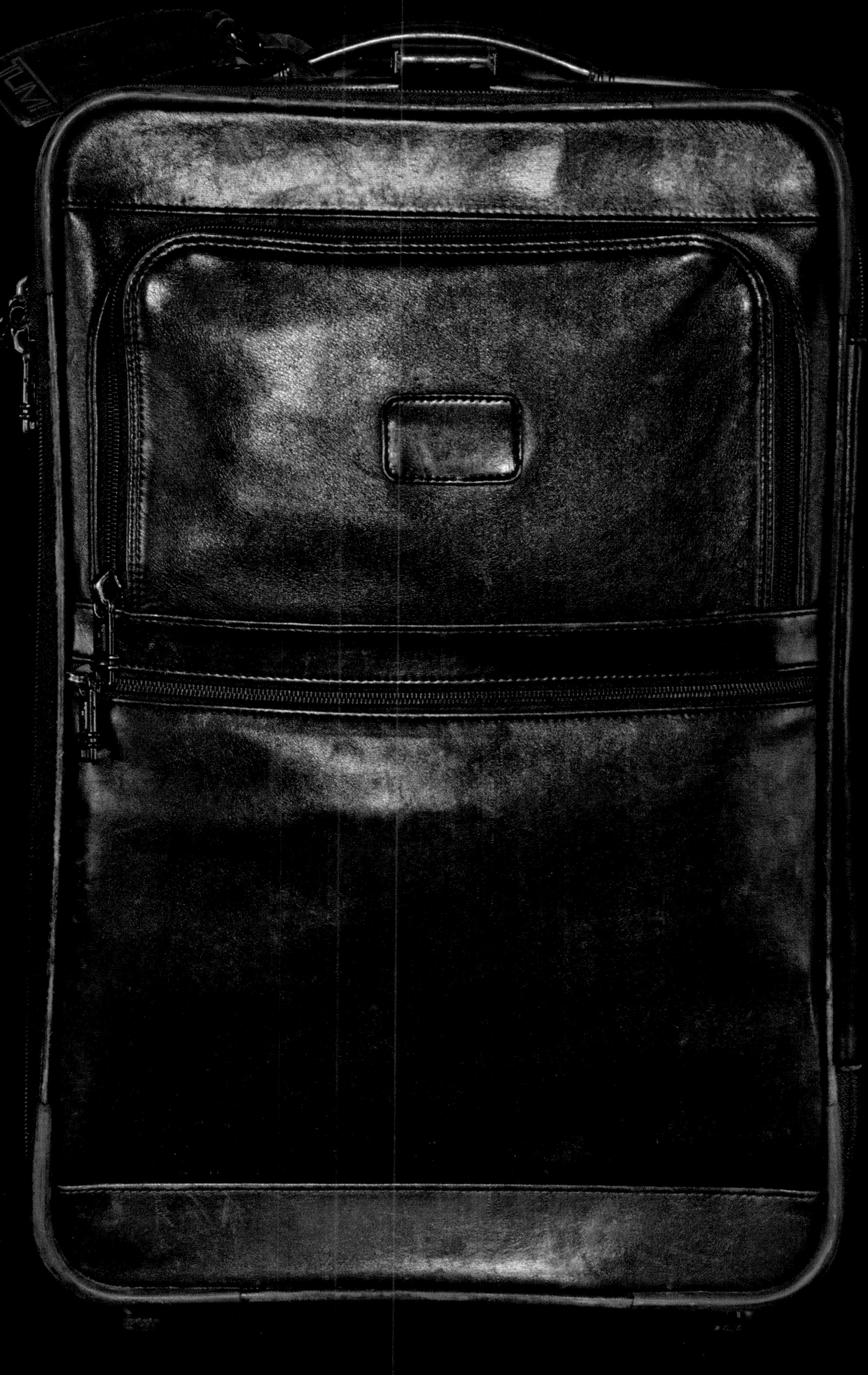

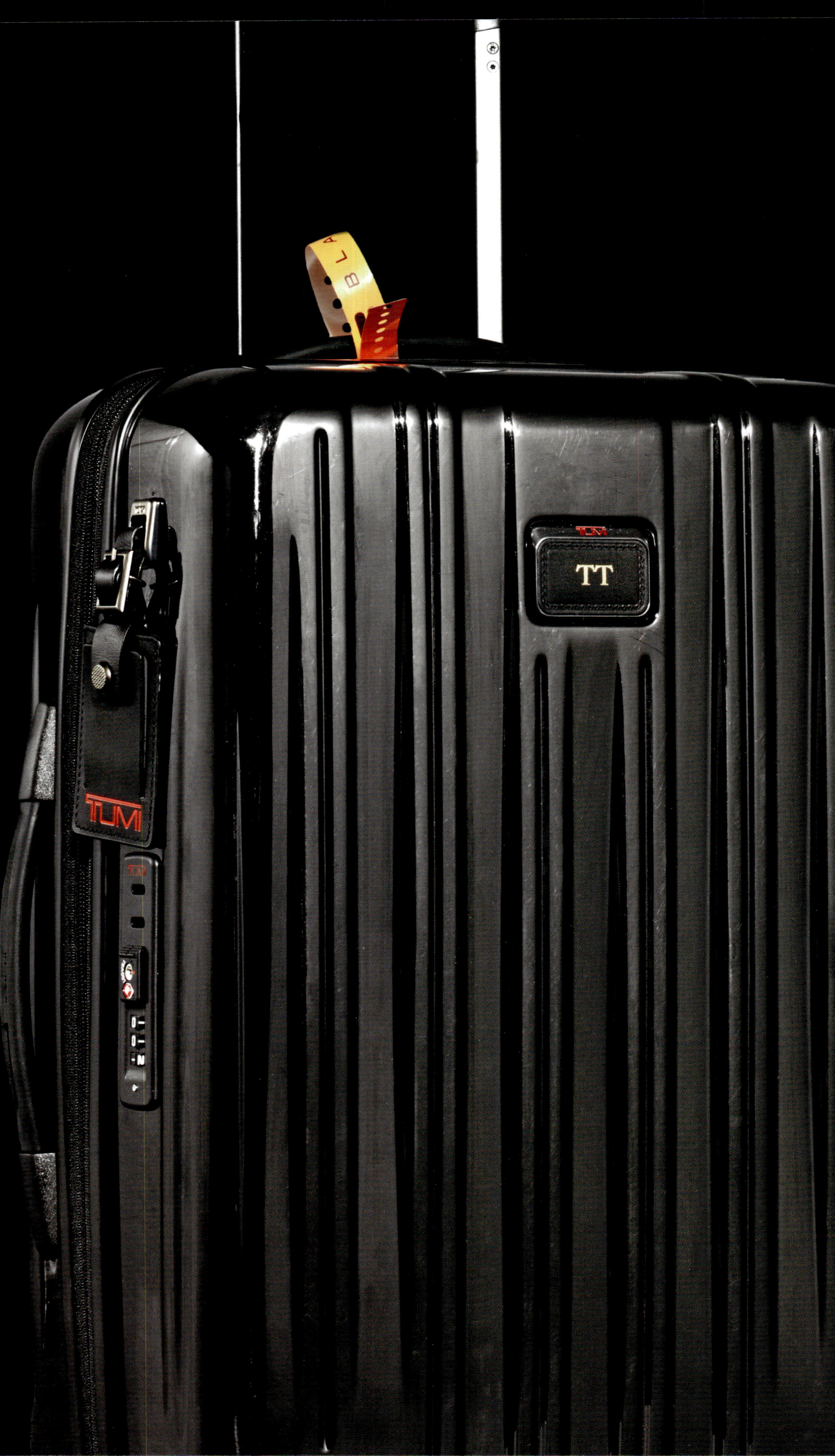
TT
TUMI

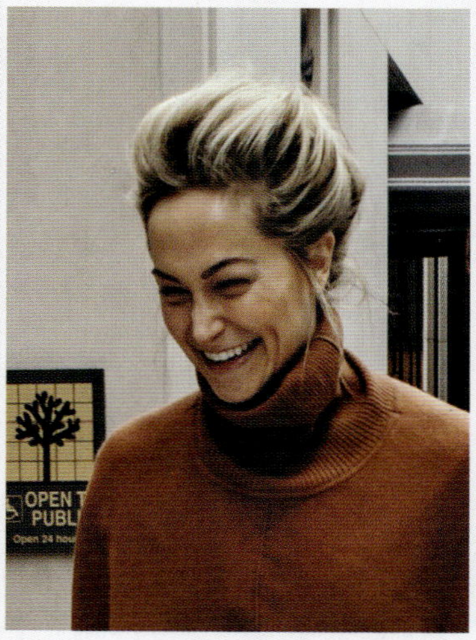

Taryn Toomey

Founder & CEO of The Class

I am the founder of a health and wellness company that supports people's own evolution. My TUMI helps me to stay organized and allows me to feel grounded when I travel.

When I first moved to New York 18 years ago, people would always say, about my name, "Toomey, like the luggage?" And I had to Google TUMI because I wasn't as savvy as I am now. But after that, when people would ask, I'd say, 'No, but I wish.' Now I absolutely live with my TUMI carry on. I can fit everything in there. I never check a bag. Not even when I travel with my two children. It's some weird magic that I do. The heavy things—scarves, leg warmers, boots, a hat—they live on my body. I recently went to Greece, Italy and London with a friend, and she said, "We'll have to check a bag," and I thought, "No way. It'll get lost on a trip like that." But I did it anyway, and checked a huge carry on and I wound up wearing three things out of it. It was ridiculous. I do have a uniform when I travel. I wear the same four things, and if I need it, I can always do laundry at the hotel. TUMI makes me feel organized.

Collection

TUMI V3

Style

International Carry-On

Collection Launched

2016

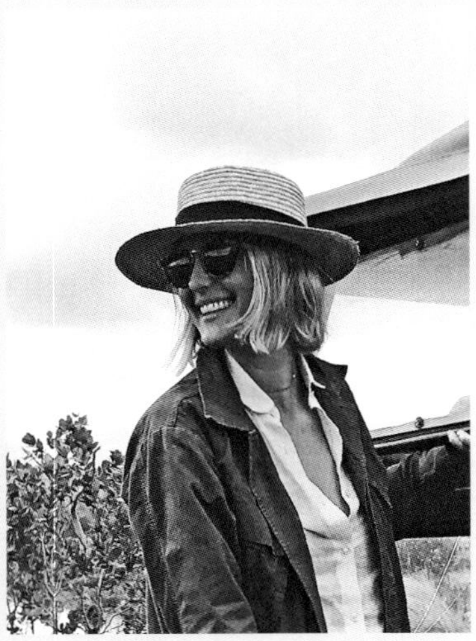

Anita Patrickson

Celebrity Stylist and Founder & CEO of Amanu

Collection

Santa Monica

Style

Whitfield Helemet Bag

Collection Launched

2013

I don't want to buy a new bag every two years. It's wasteful and excessive.

For someone who works in fashion, I am a pretty lean packer. I hate getting somewhere and feeling like I have to spend my time making decisions on what to wear or piecing together specific outfits. I just want to be present and focus on wherever I have traveled to. I own a custom sandal company, Amanu. We design beautiful, bespoke sandals that pack 'flat', go with everything and travel beautifully. My other secret travel packing tip are my wraps and kikois which you can find all up and down the East coast of Africa, especially on the beaches of Somalia and Kenya where my family is from. Everyone wears them to the beach in Kenya. They're great. I use them as scarves or wraps—I can even carry things in them. The more you wear them the better they get. I don't like things to feel too precious. That's why I love my crossbody TUMI bag—the more I travel, the better it gets. I've carried this bag for almost ten years. Luggage should be really special and beautifully crafted, not something to cast aside. You should be able to wear it forever.

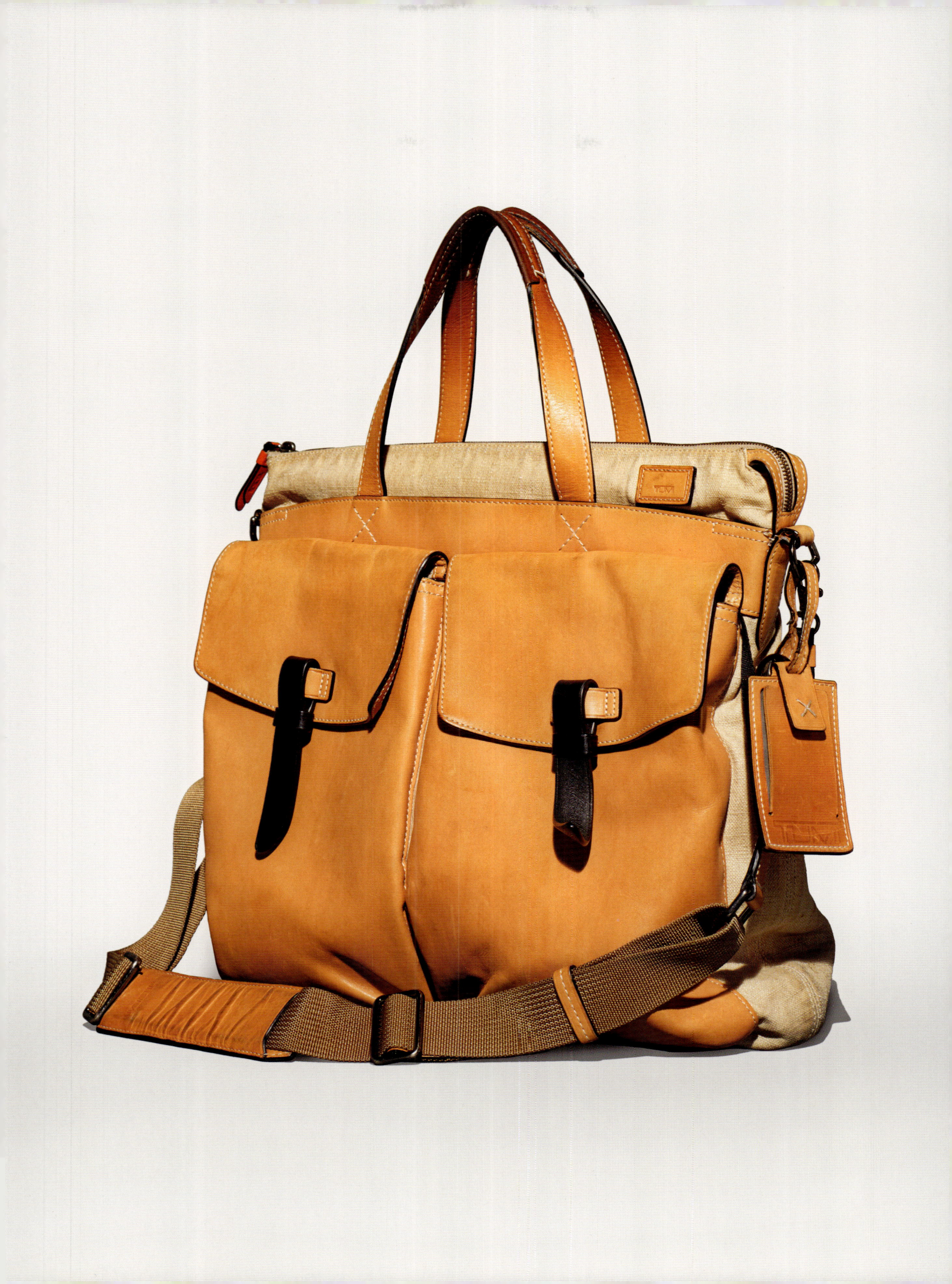

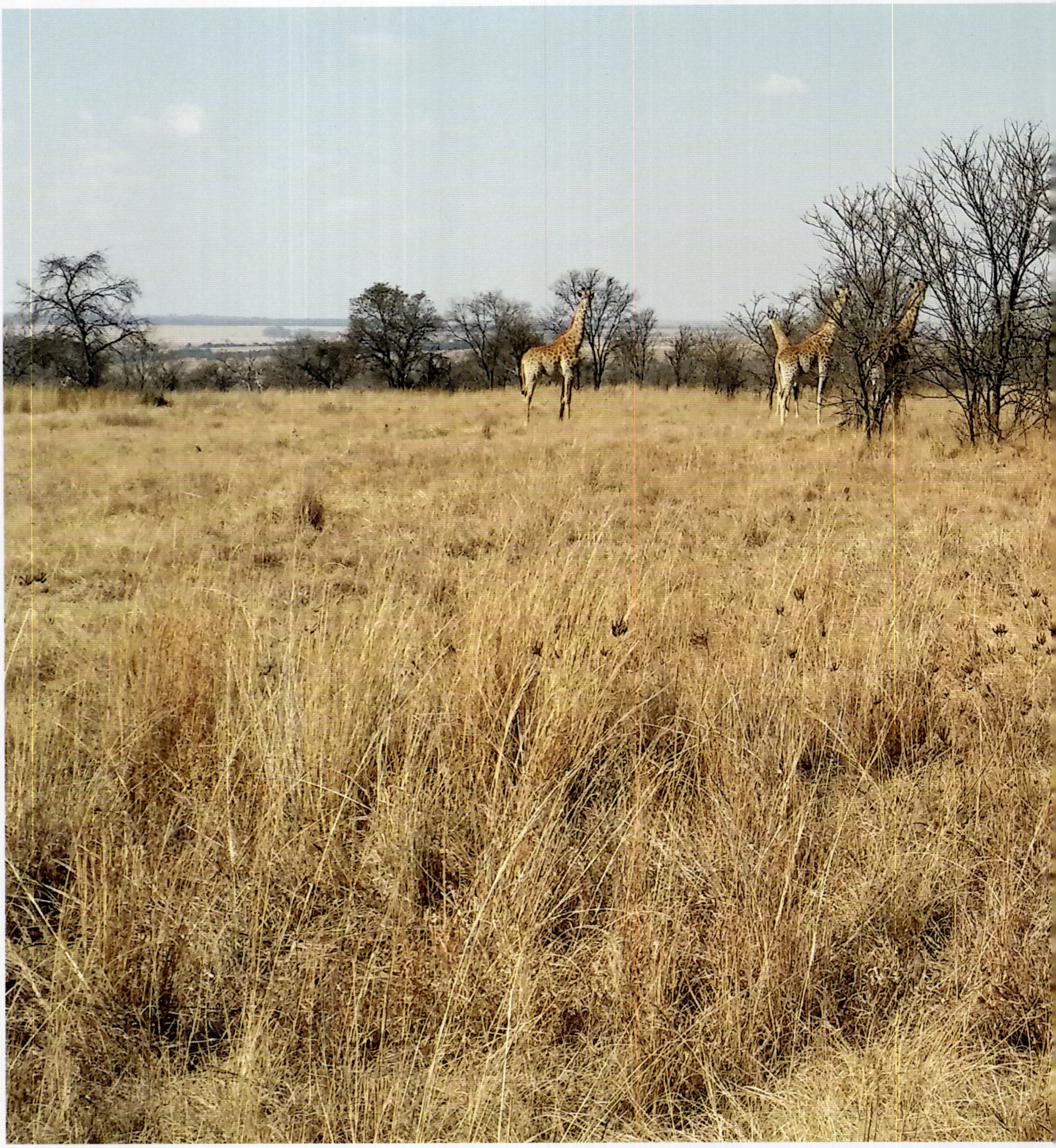

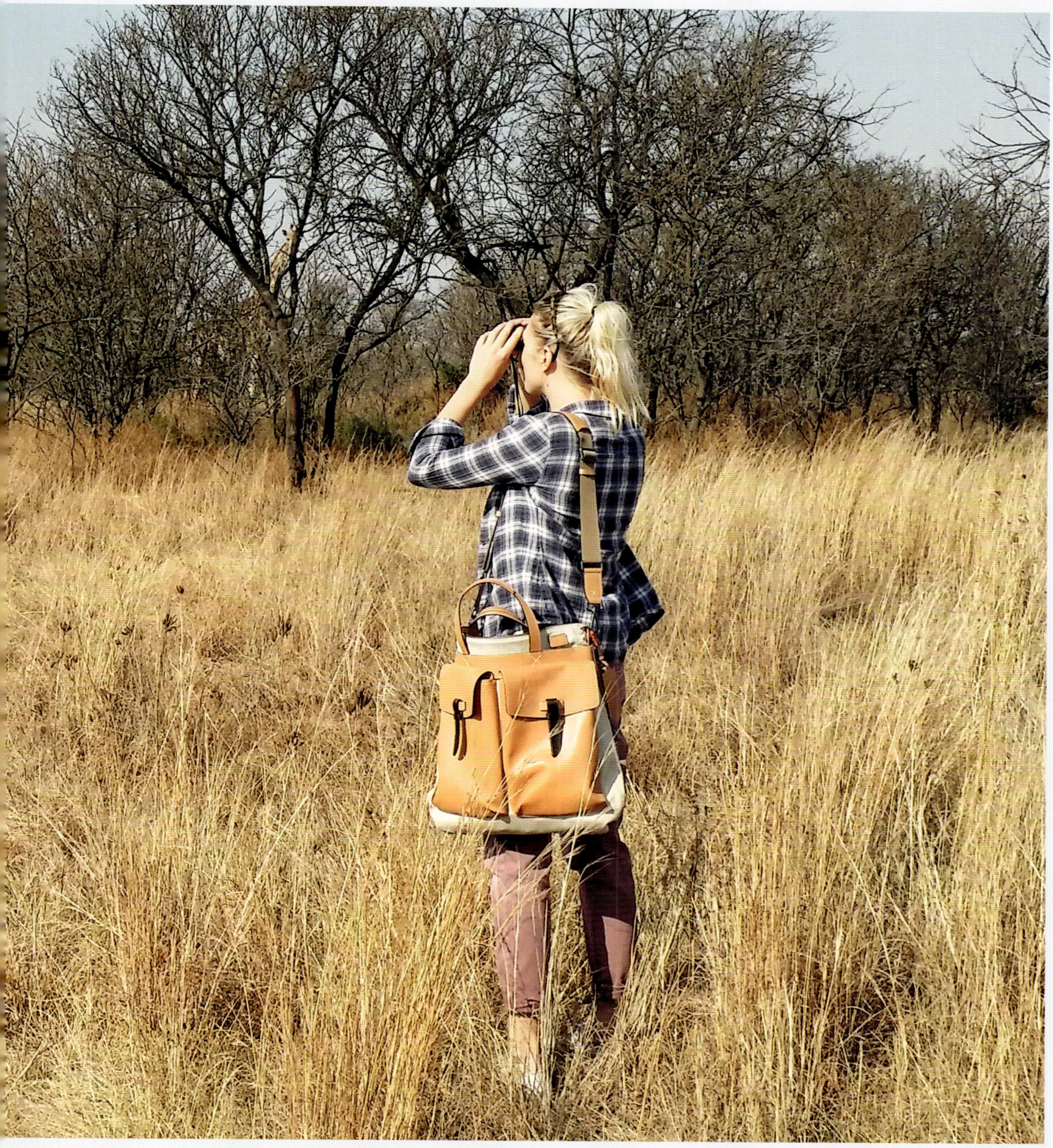

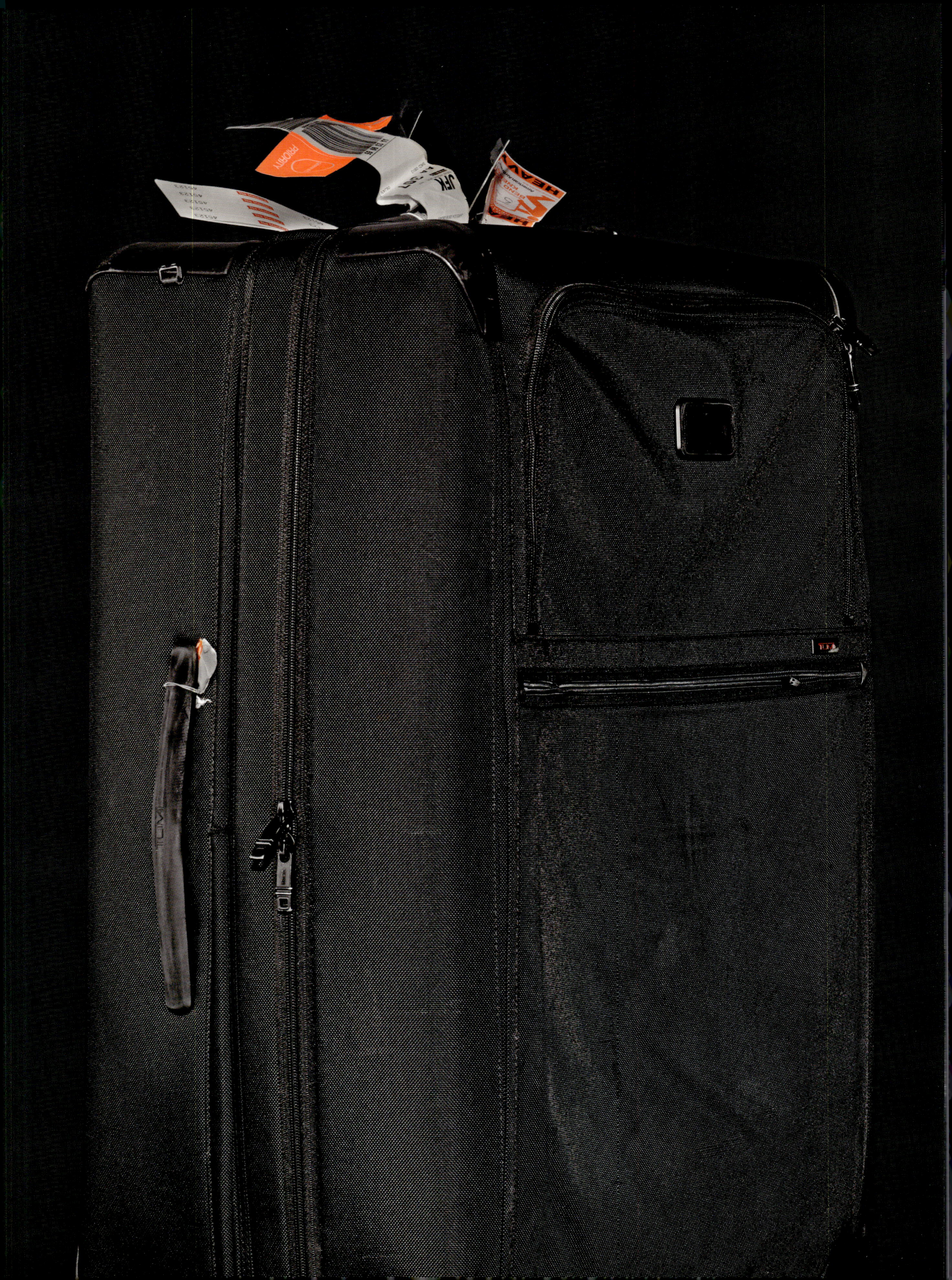
JFK
HEAVY

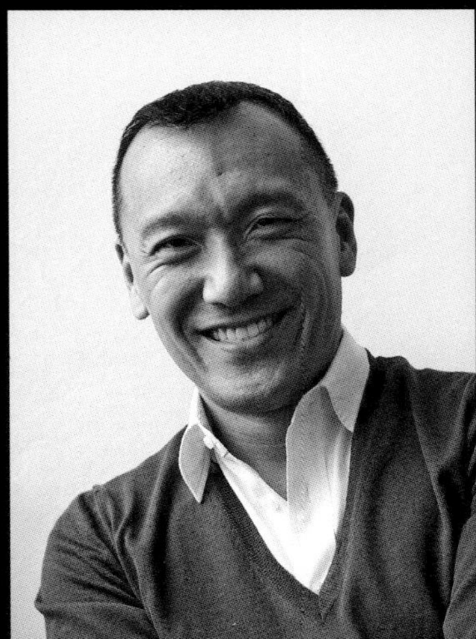

It's important that my luggage is simple and classic. And I work in fashion, so it's got to be black.

Collection
Alpha 2

Style
Worldwide
Packing Case

Collection
Launched
2016

Joe Zee

Fashion Journalist and TV Producer

I learned about TUMI when I was a fashion assistant. I was always on a plane, and I had to carry so many clothes and accessories. And those things had to be packed a certain way, which meant I needed something sturdy. But my luggage also had to be elegant and chic—with no screaming logos or initials. I used my first TUMI for about 12 or 15 years. It went everywhere with me. About five years ago, the zipper wasn't working properly, so I took it in to be fixed but wound up buying a new bag. I haven't thrown out my original though. I've had it for so long, I just couldn't part with it. TUMI affords me protection. I know nothing will be damaged inside my suitcase. I rely on that. It's also important that my luggage is simple and classic. And I work in fashion, so it's got to be black.

Deborah Lippmann

Brand Founder and Celebrity Manicurist

Even when I couldn't afford a TUMI, I found a way to get one.

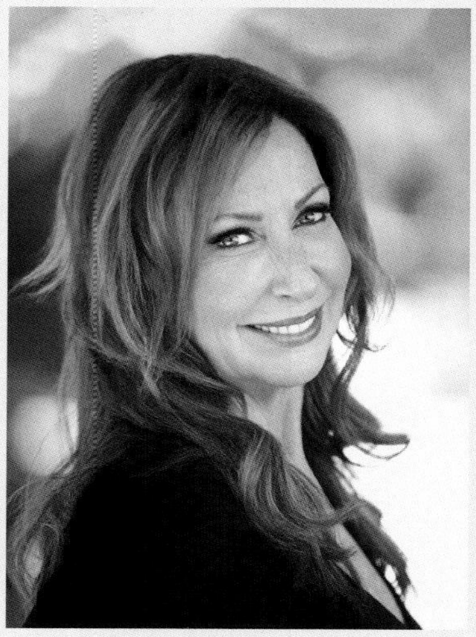

I bought my first TUMI bag in 1995, and in the years since, I've probably bought 50 or so pieces. I first saw them on set: makeup artists and hairdressers all said they were indestructible, and at that time, I was buying luggage from a place on 14th Street in Manhattan. Those pieces couldn't handle a curb and would break in 30 days. What a waste of money.

I'm a schlepper. And because I'm carrying these tiny glass bottles, I need a bag that's really secure. I have a huge duffel that zips at the bottom that I call The Gaga Bag. I use it when I'm with a client who is going to want really progressive nail art, like Lady Gaga. I'll pack fabrics and supplies from M&J Trimmings in New York and different kinds of nail tips and machines, like drills—you have to bring every color because you never know what's going to happen. The Gaga is the largest one on wheels. I can actually roll it and it doesn't tip over. And if you take it down a curb, it remains upright. If you travel all the time, you understand.

Collection
T-Tech Pulse

Style
2 Wheeled Duffel

Collection Launched
2005

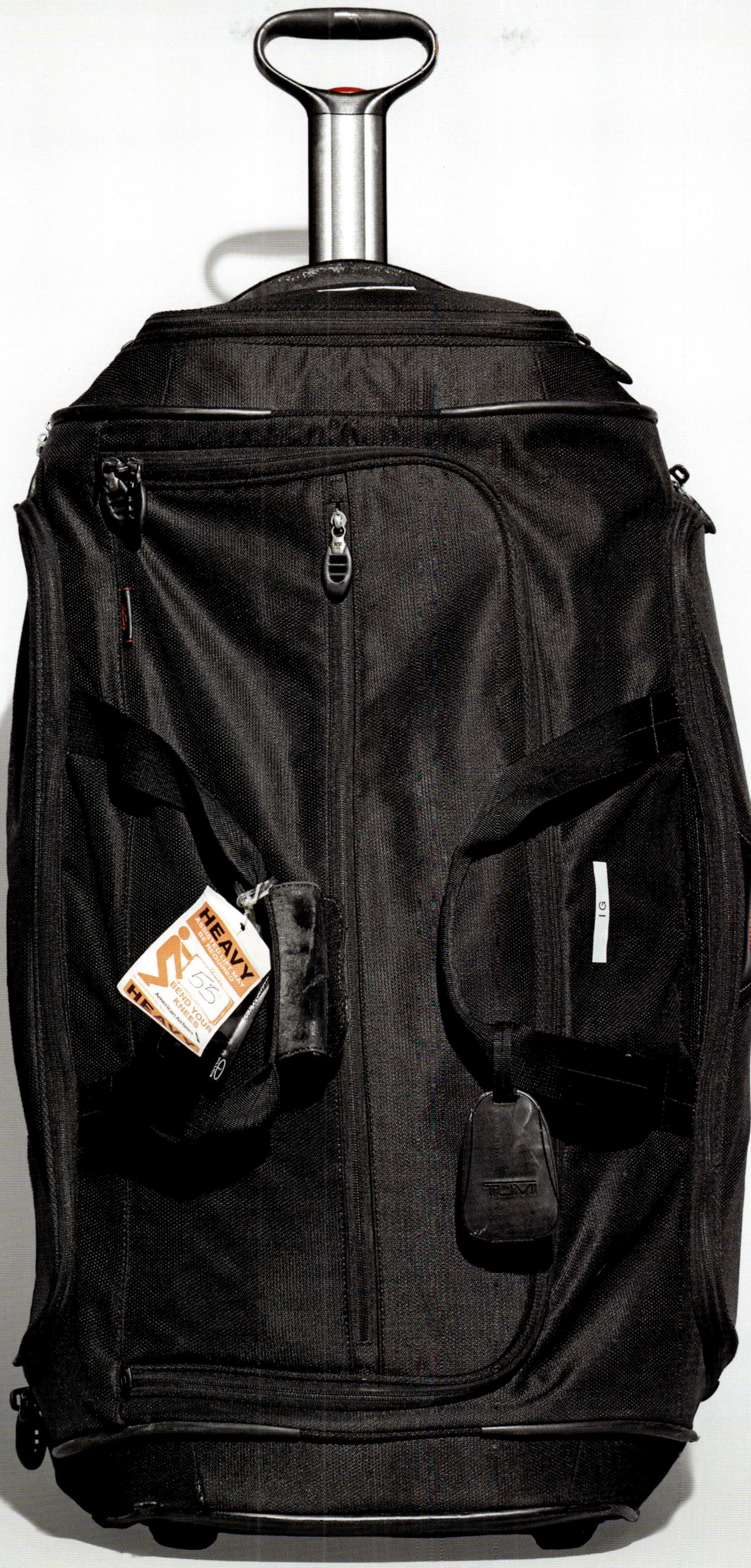
HEAVY
53
BEND YOUR KNEES
HEAVY

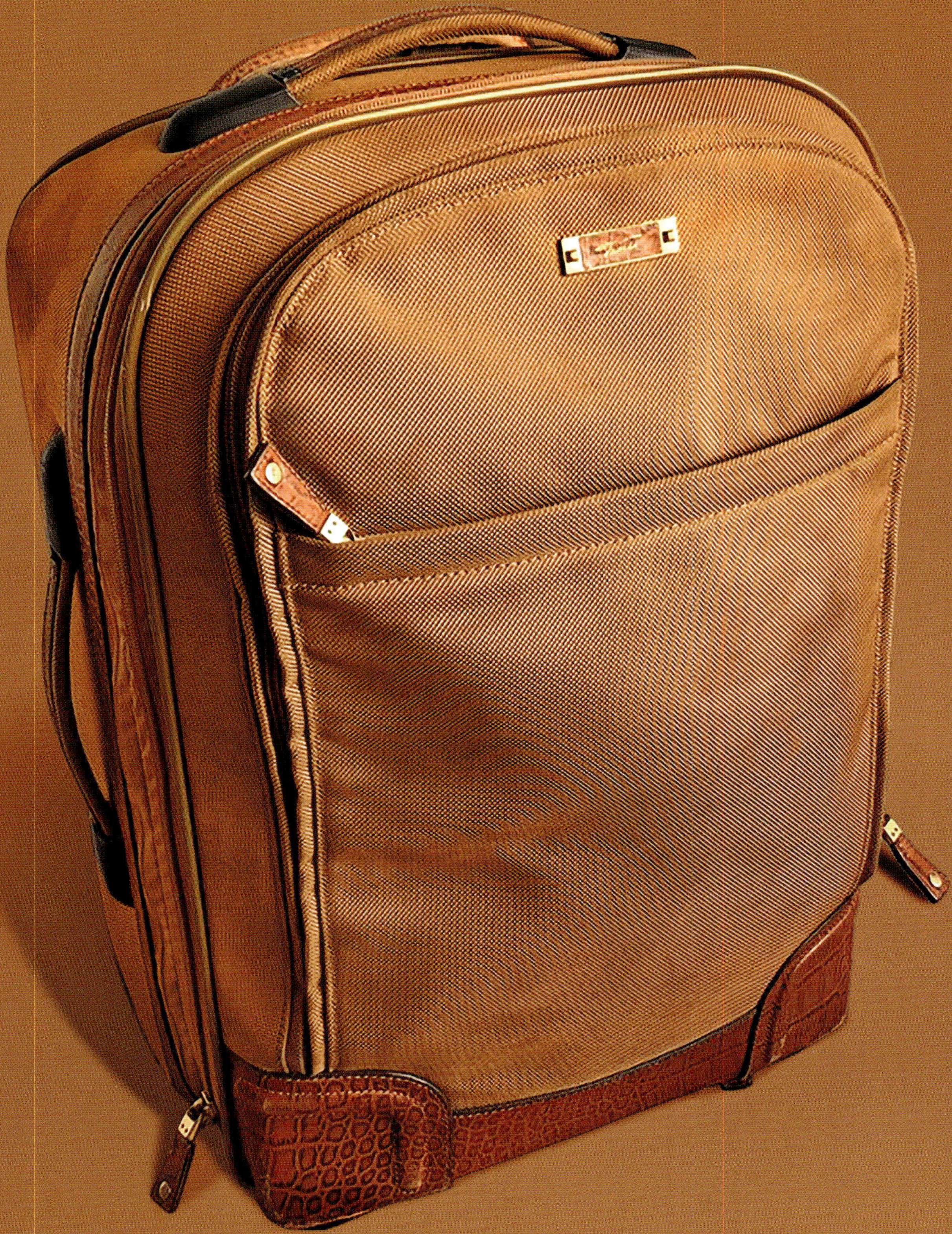

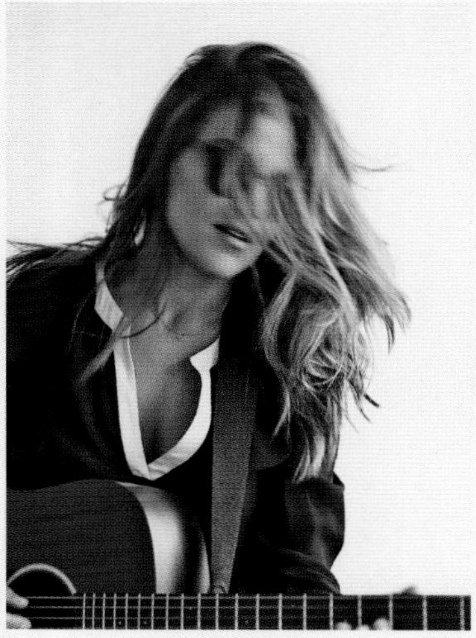

Collection
Georgetown

Style
Potomac
International
Carry-On

Collection
Launched
2012

Melody Gardot

Author, Composer, and Singer

I discovered the brand about ten years ago. I found out about it from a director friend of mine who was traveling the globe to shoot music videos and commercials. When we met, he had his luggage with him. I noticed it and exclaimed, “That case is amazing. It’s huge! Where’d you get it?” And he said it was TUMI. Once I got back home I ordered one . . . then another. . . and another . . . Before I knew it, I had four!

Because I’ve been touring for so long, I’ve really “lived” out of these bags for the last 10+ years. There was even a time when we were working with Lanvin and Alber Elbaz was dressing me for my shows. I was carting around all the garments in two of my TUMI cases as I didn’t have a wardrobe case. The dresses were so delicate, but my TUMI had a three fold envelope inside top lid for really long dresses. The bottom was full of boxes holding shoes and jewelry for the stage.

The design could not have been better during that period of time as we had so much to carry and it all needed to be kept immaculate. There were side pockets on either side that were the length of the suitcase. I just pulled two satin belts out of one of the pockets and they’re perfect. Mind you, they’ve been in there for almost four years. Impressive!

When I first started working, I thought dropping $800 on a bag was insane, but I did it because TUMI has a great lifetime guarantee, which means you’ll have this bag forever. I thought of it as an investment. As long as I’m alive, this bag is going to work.

I’ve been touring for so long, I’ve really “lived” out of these bags.

My Just in Case bag is sort of like the Mary Poppins bag.

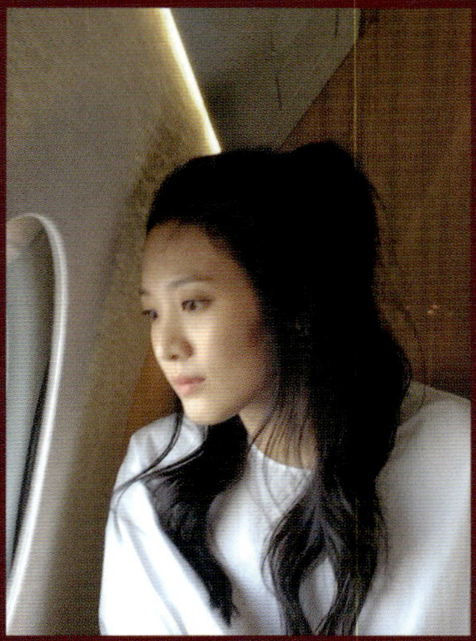

Claudia Kim

Actress and Model

I'm always packing and unpacking. Normally I travel with three big travel cases and one carry-on because when we're filming, I'm gone for six to eight months at a time. I always always have my Just in Case tote. I can fit so much in there—my tea, sneakers, an extra change of clothes, snacks, protein shakes, and at least two TUMI makeup pouches. I carry a lot of tea. For some reason, I always feel like I'm going to run out of Twinings breakfast tea, so I make sure to leave room for extra boxes in my carry-on. It's like a magic bag. It's totally replaced my purse. And it's on set with me all the time. Everyone who works with me knows that.

Collection

Voyageur

Style

Just In Case Tote

Collection Launched

2018

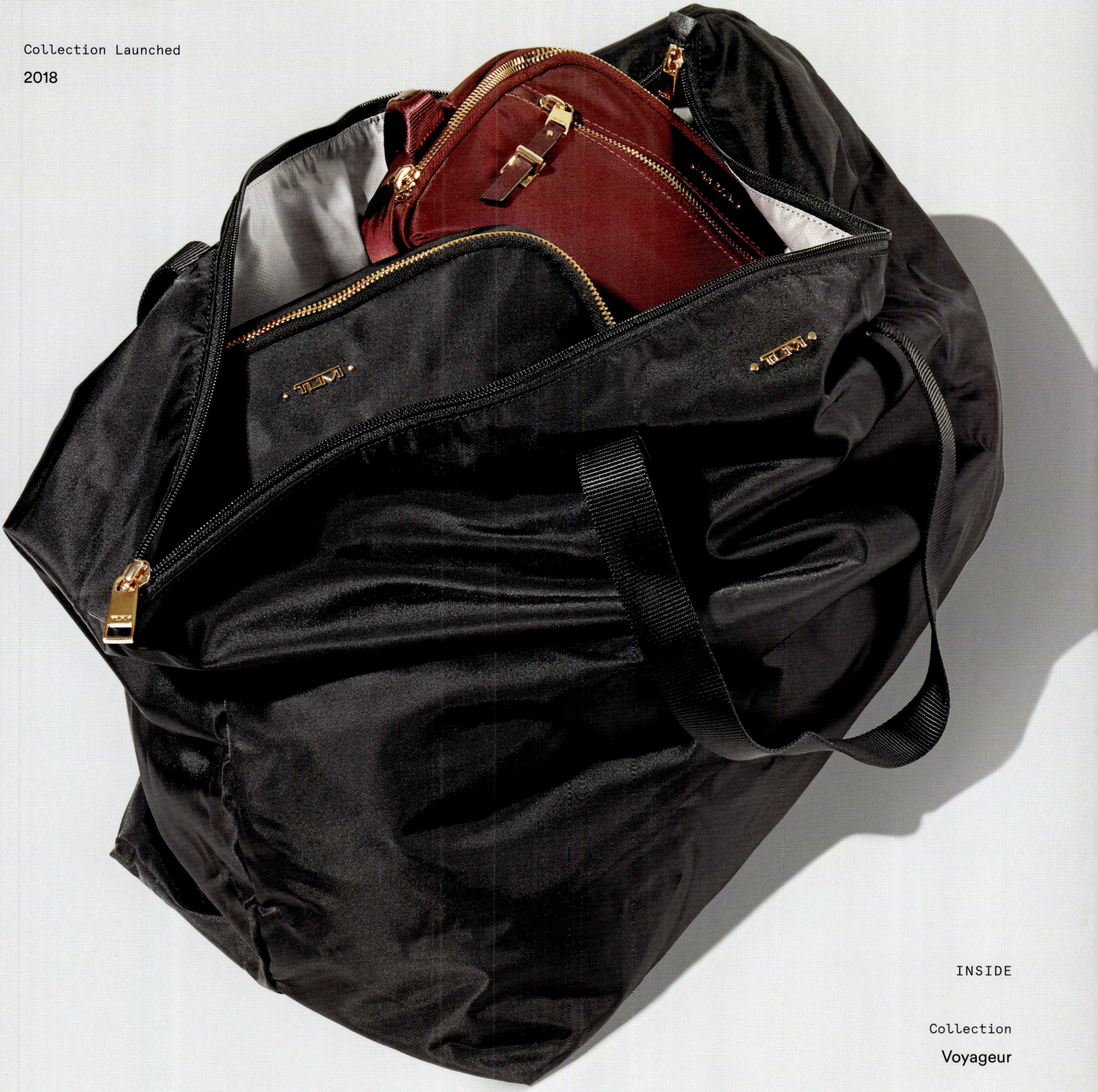

INSIDE

Collection

Voyageur

Style

Madina Cosmetic Cases

Collection Launched

2018

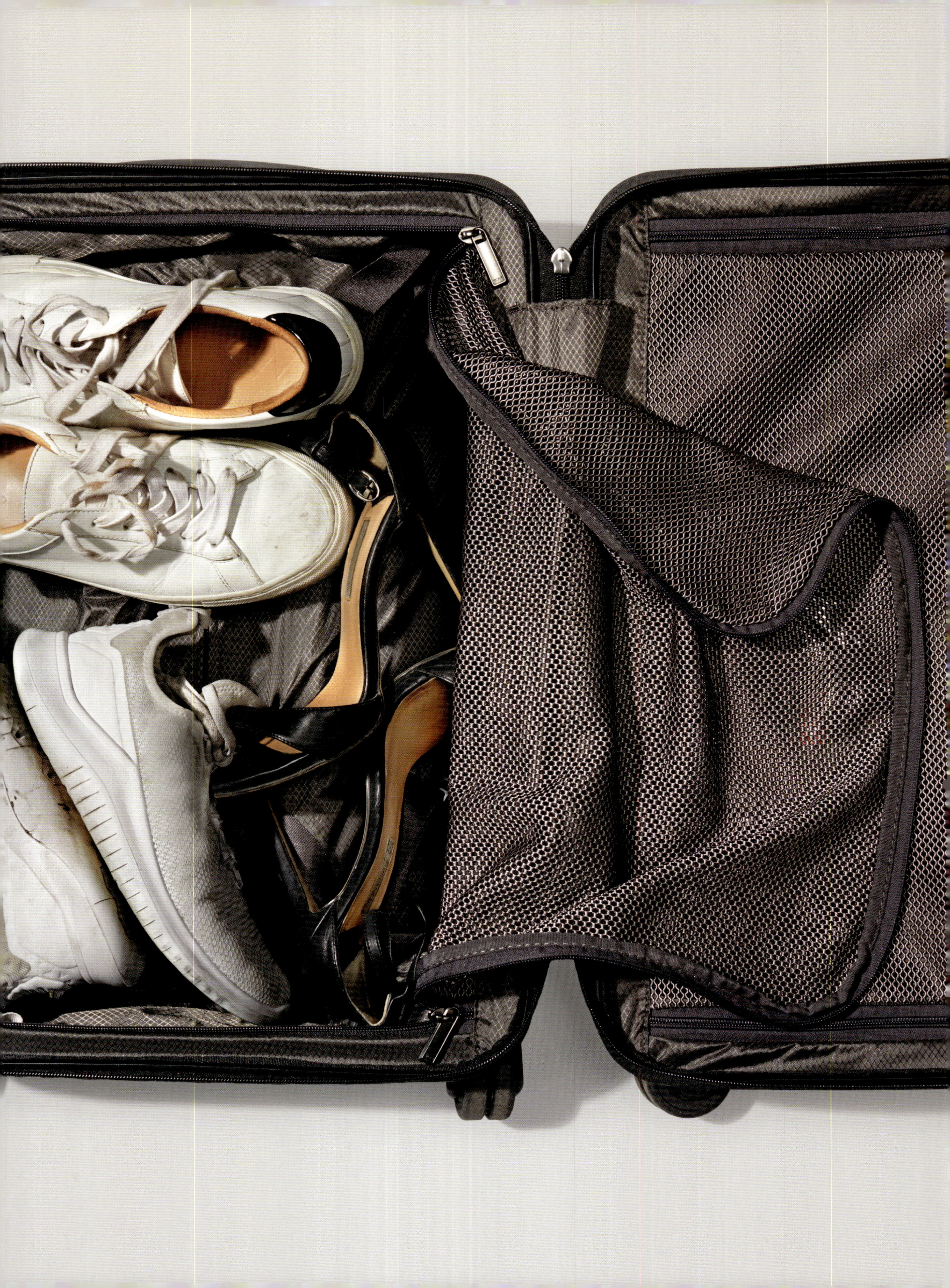

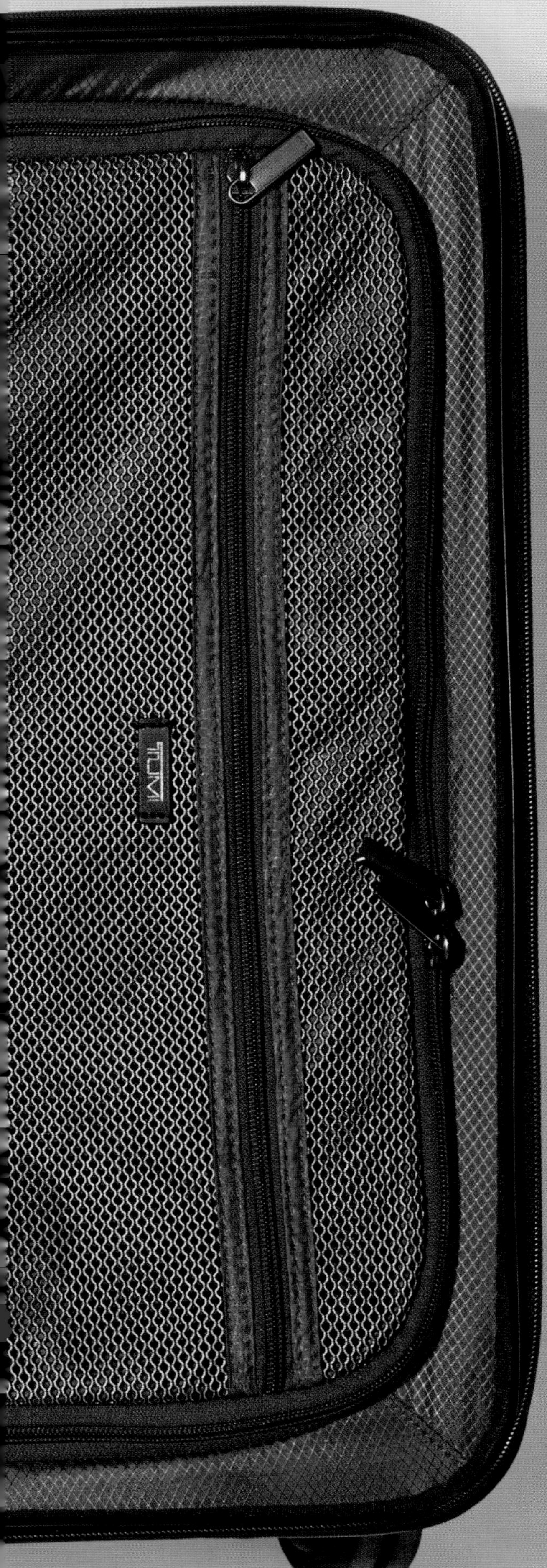

I like to keep things simple.

Charlotte McKinney

Actress and Model

When you think travel and luggage, you think TUMI. It just stands out for being a classic. I don't think it's necessary for me to make a fashion statement with my luggage. But I'm always trying to pack as much as I can into one piece—especially my tennis shoes and my heels. Those are the biggest things I'm always trying to cram into my carry-on. Knowing I have my bag with me grounds me in a way. Other bags really do come and go—some charge your phone, others have crazy patterns. But TUMI is timeless.

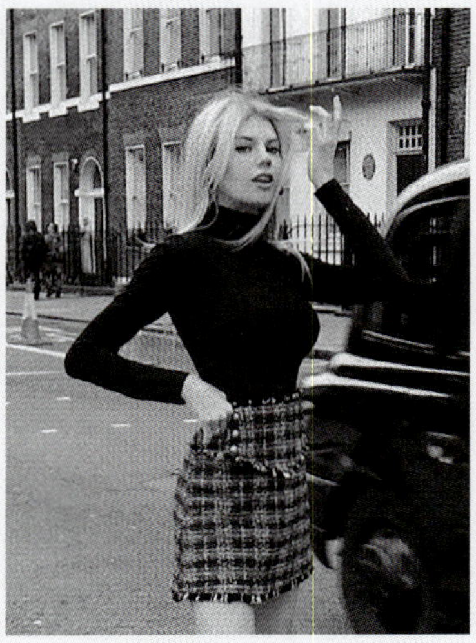

Collection

19 Degree PC

Style

Continental Carry-On

Collection Launched

2016

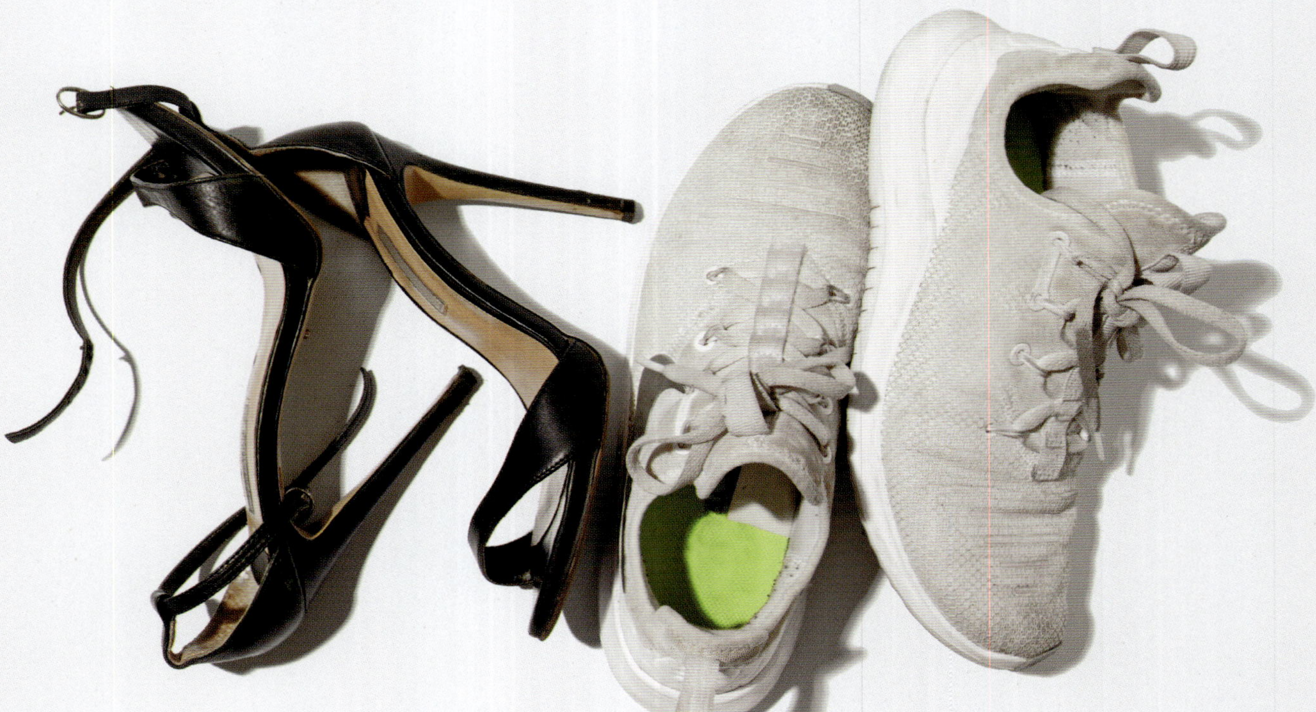

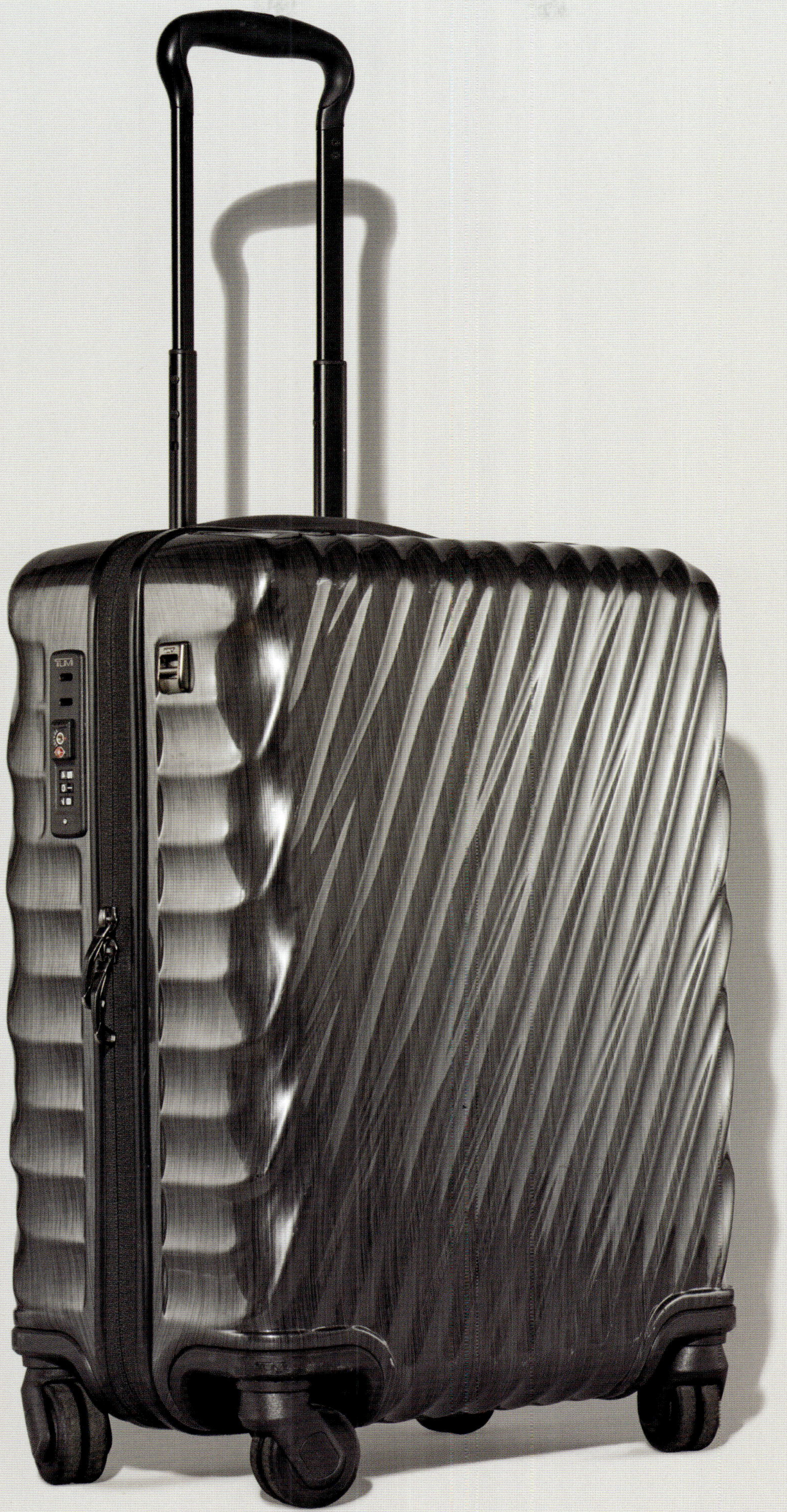

It never crossed my mind until that moment that people used nice luggage.

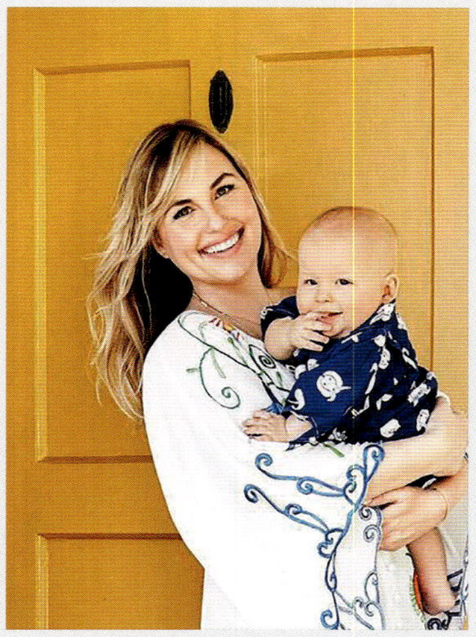

TOP

Collection
Gen 3

Style
Weekend Tote

Collection Launched
1993

Hillary Kerr

Co-Founder and Chief Content Offier, Clique Media

I graduated college a semester early so I could move to Sydney for six months for internship at *Marie Claire* magazine. I was very excited and feeling a bit like a grown up. My parents threw me a small graduation party at their house and invited a group of friends I've had since kindergarten and their parents. They all chipped in and bought me a TUMI passport holder and a really beautiful black leather carry-on bag. I'd never seen such luxurious luggage in my entire life. My dad is an airline pilot and former military pilot, so I was used to the all-purpose suitcases we'd buy at the military exchange depot. Those bags weren't an extension of you. They were purely functional. So to have a bag at barely 21 years old that was so elegant and understated, yet so beautifully made, was just astonishing to me.

I'm 39 years old now and it's still my carry-on. I use it a few times per month. I've taken that bag all over the world and it looks pretty much just as it did when I got it. It was with me on a work trip I took to New York while I was pregnant with my son. I wound up giving birth to him on that trip.

In fact, when TUMI heard that I was having my first child unexpectedly in New York, when I was planning to have him back home in Los Angeles, they were so incredibly kind and sent Clark his own set of TUMI luggage. Which we desperately needed because we had to take so much gear with us to drive back home. I still can't believe the first time I drove across country was with my dad, husband and six-day-old son. It was a very wild ride.

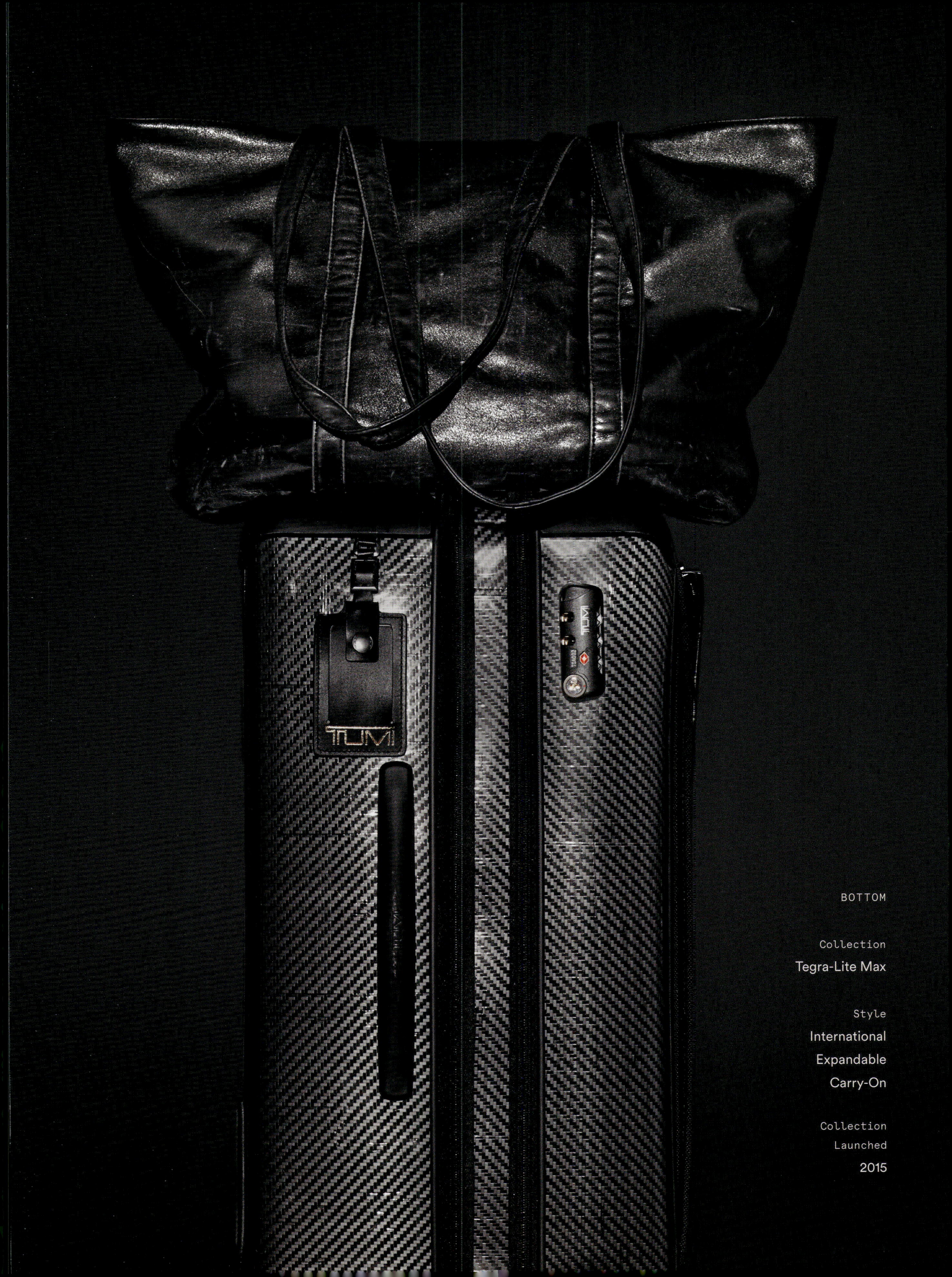

BOTTOM

Collection
Tegra-Lite Max

Style
International
Expandable
Carry-On

Collection
Launched
2015

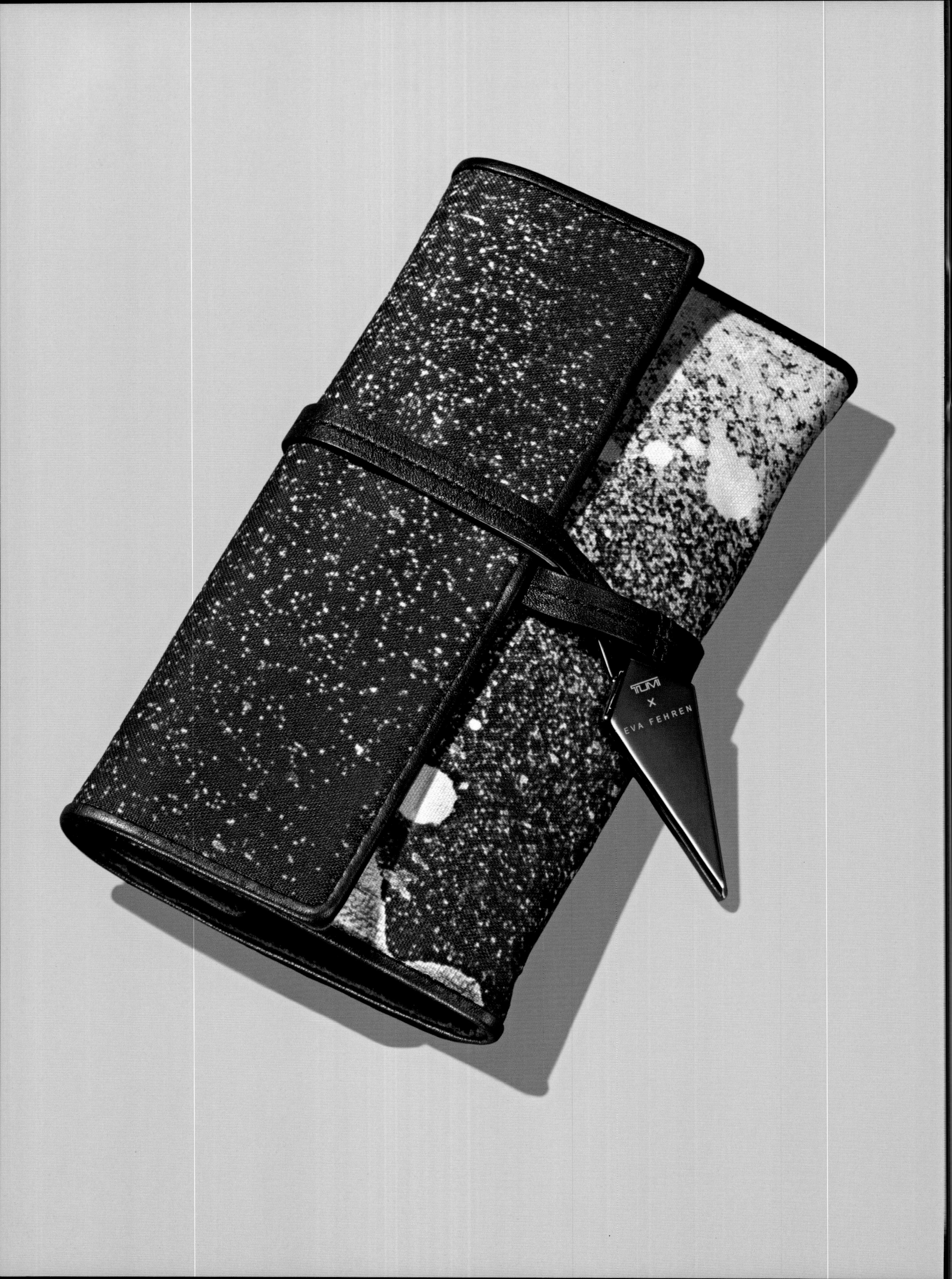
TUMI
X
EVA FEHREN

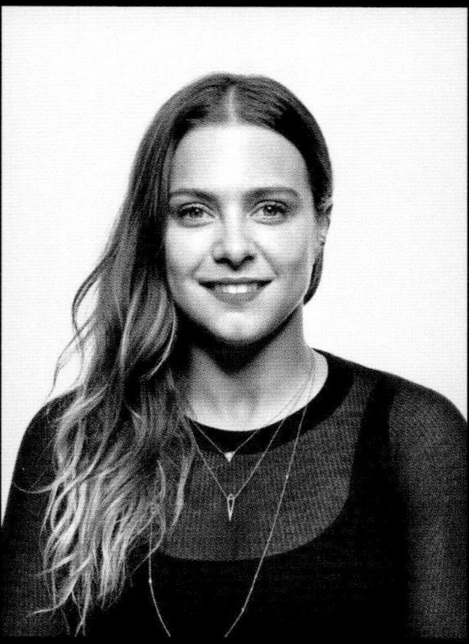

Collection

TUMI x Eva Fehren

Style

Jewelry Roll

Collection Launched

2017

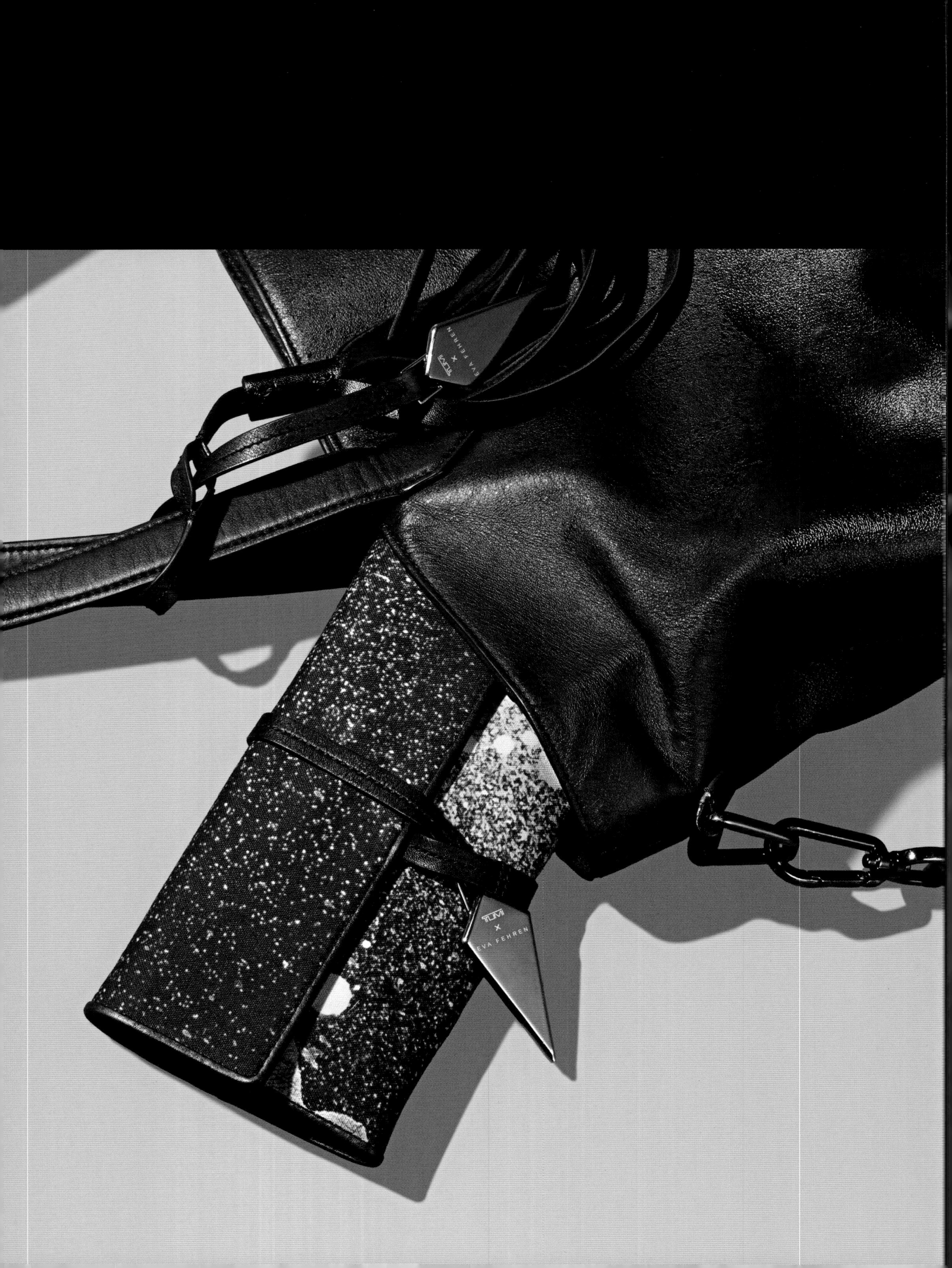
TUMI
X
EVA FEHREN

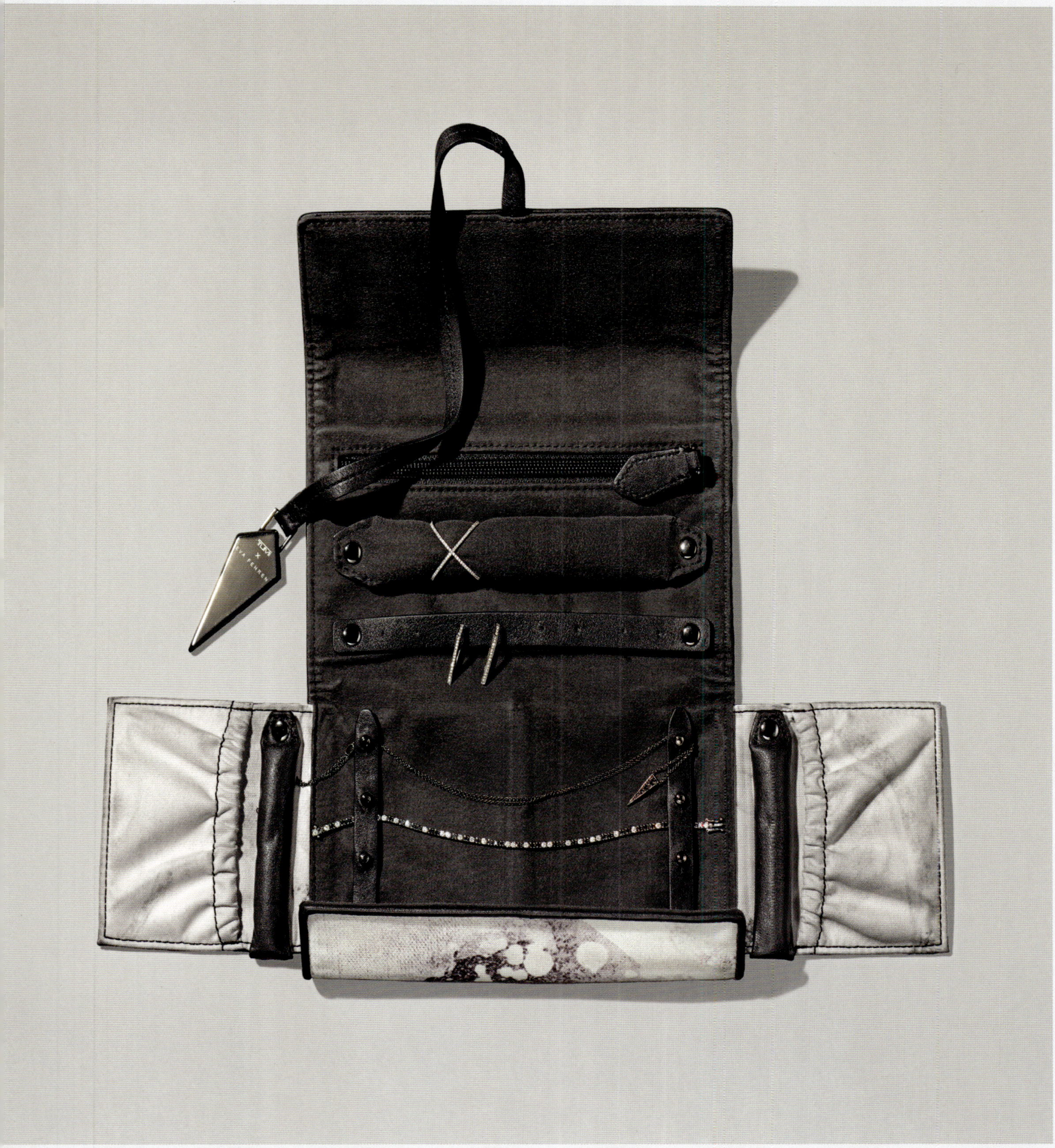

Kathryn Minshew

CEO and Co-Founder of The Muse

If I have to check a bag that means I am doing something wrong.

When I joined McKinsey & Company as a consultant in 2008, it became very clear that all the consultants had TUMI luggage. I was young and new to the workforce, but all these people who I respected trusted the brand. I knew they couldn't be wrong. That's when I got my first TUMI—a black roller carry-on. I quickly became a seasoned road warrior. I was living in New York City, leaving on Monday mornings and flying back on Thursday afternoons. My bag has been beaten up, but I still use it for all my travels. Now I travel three to four times per month—75 percent of that is for work. I love travel, but right now it's mostly for business. At some point though I'll make sure to get back out there on the road, for pleasure.

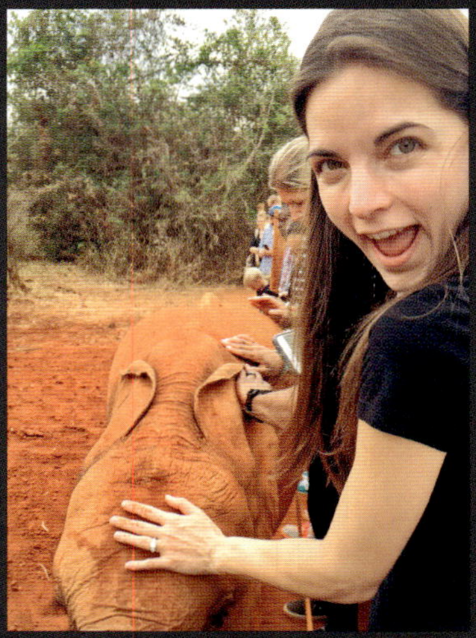

Collection
T-Tech Move

Style
Overhead Carry-On

Collection Launched
2007

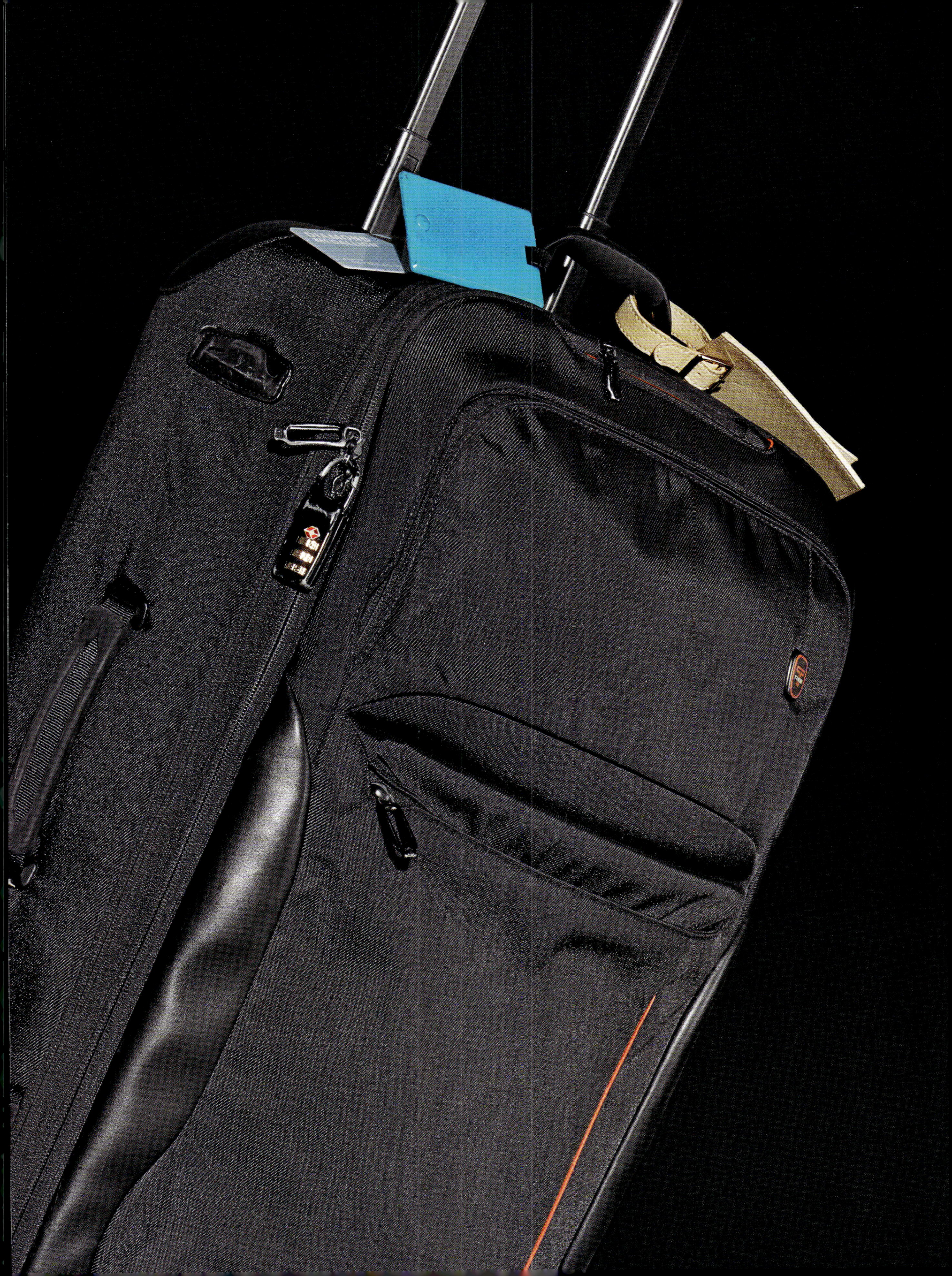

LEFT

Collection

Voyageur

Style

Monaco
Cosmetic Case

Collection
Launched

2018

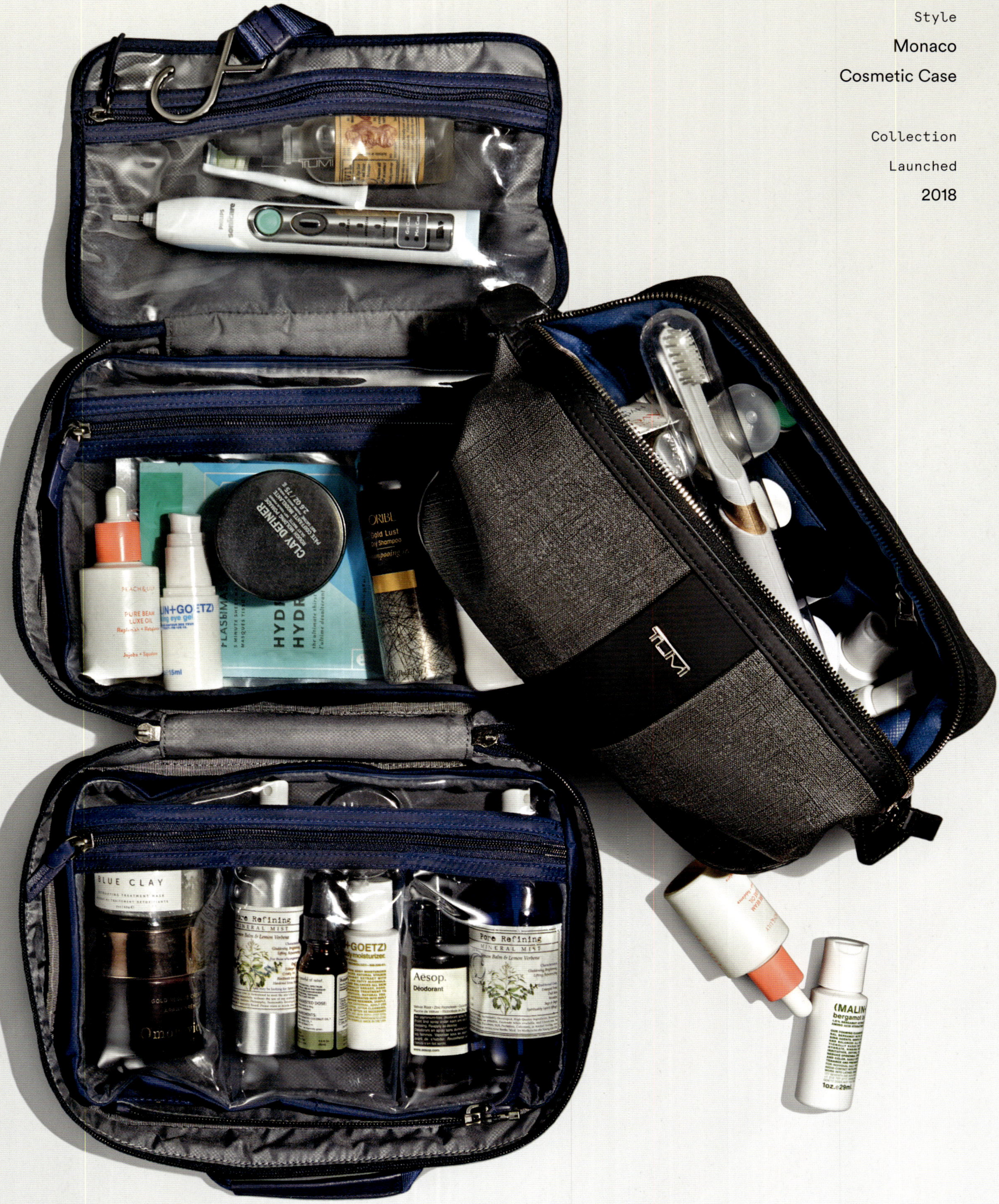

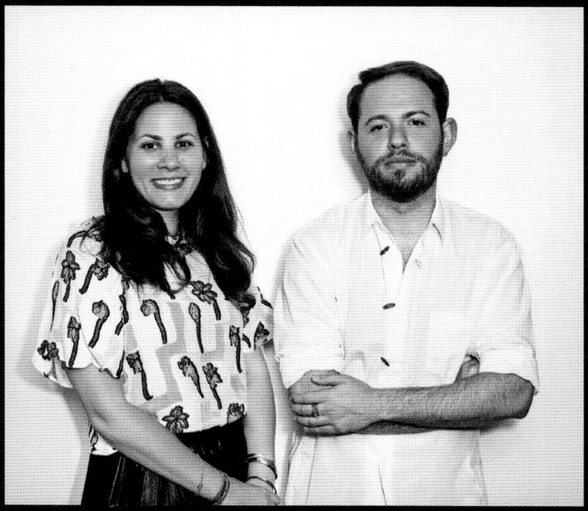

RIGHT

Collection

Ashton

Style

Raymond Travel Kit

Collection Launched

2018

Jonathan Cohen & Sarah Leff

Co-Founders of Luxury Ready-to-Wear Label, Jonathan Cohen

Once you have a TUMI you can't switch back to anything else.

Sarah: Growing up, I had TUMI pieces here and there, but in the last four years I've turned everything over to TUMI. The pieces really do make travel more convenient and enjoyable. With my Dopp kit, I never have to worry about anything breaking or leaking. I keep it stocked with face and eye masks and lotions. It's always ready to go. In my carry-on, I'll pack a pouch with an extra change of clothes—always, just in case—my TUMI adapter and a notebook and pen. I also have a TUMI roller bag. It just glides through the airport. Once you have a TUMI, you can't switch back to anything else. You move so seamlessly from your car to check-in. My TUMIs are my babies.

Jonathan: We travel so much that TUMI's become a natural part of my life. I have the carry-on and then the carry-on that slips over the handle of my carry-on. Which is great because it means nothing heavy on my shoulders. I love the design—it's so clean and classic. And I really love my blue nylon Dopp kit. It unfolds like an accordian, and it's completely changed the way I pack. Before I had this particular style, I'd just throw everything in toiletry bag. But my TUMI is so easy to organize. Honestly, since I got back from my trip last week, my kit has been hanging in my bathroom at home just because it's much easier to grab my things from there.

We have weathered many a storm together over the years, my suitcase and I.

Rosario Dawson

Actress and Activist

I have long since been a fan of TUMI. I often say I live on a plane, so ultimately my suitcase is my home. We have weathered many a storm together over the years and TUMI has been a great travel companion on my myriad journeys. When I first joined the campaign, I originally thought it was for luggage, then quickly became grateful to learn it was for this lovely collection. As a New Yorker, I'm used to leaving the house knowing the adventure can last all day and go in multiple directions. It's good to be prepared for anything and the versatility of this elegant and thoughtfully designed line will see you through it all.

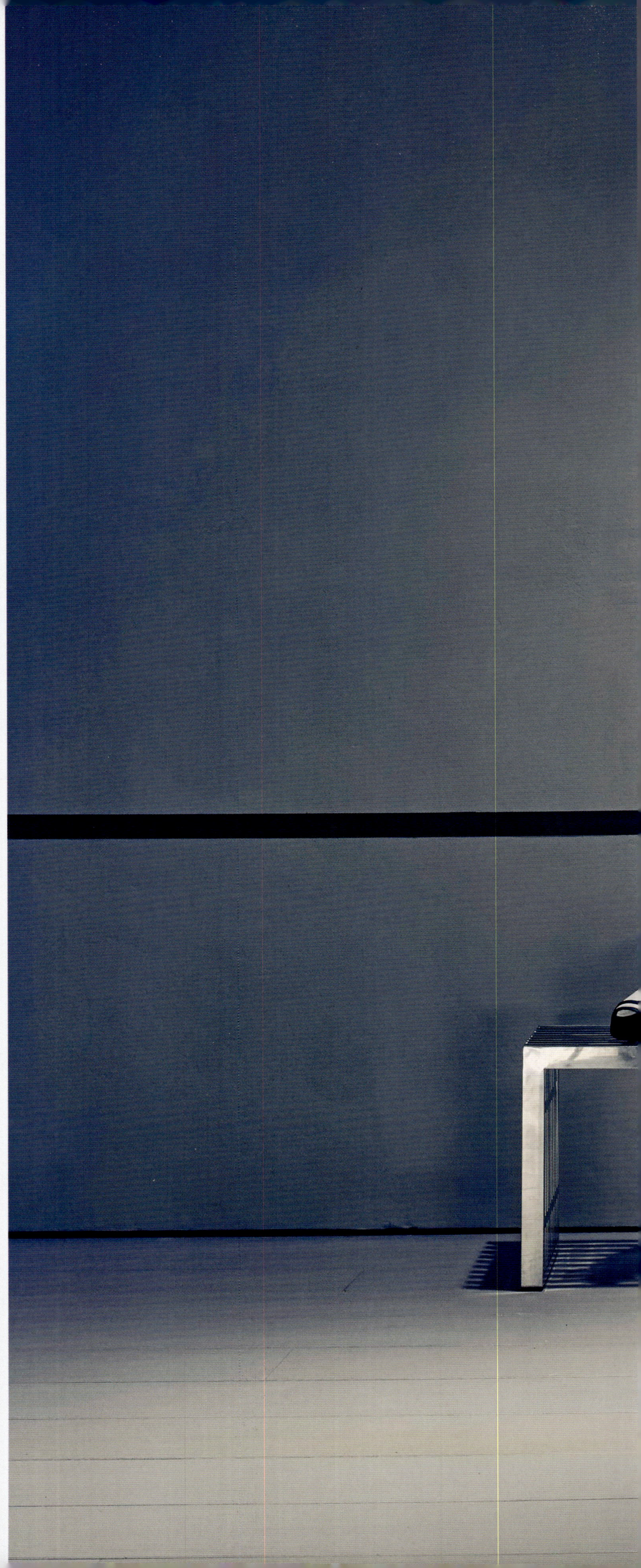

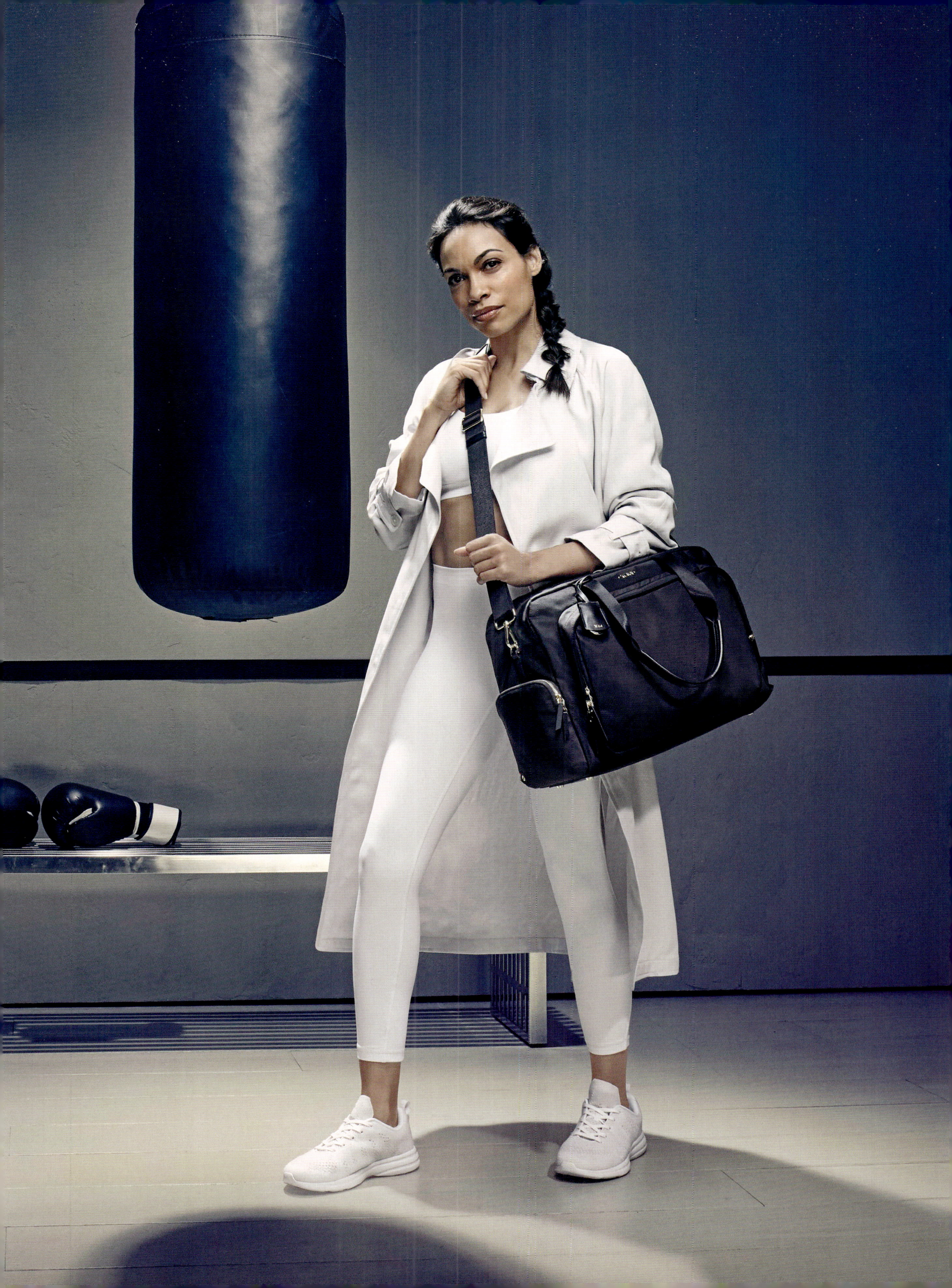

Collection

Voyageur

Style

Colina Duffel

Collection
Launched

2018

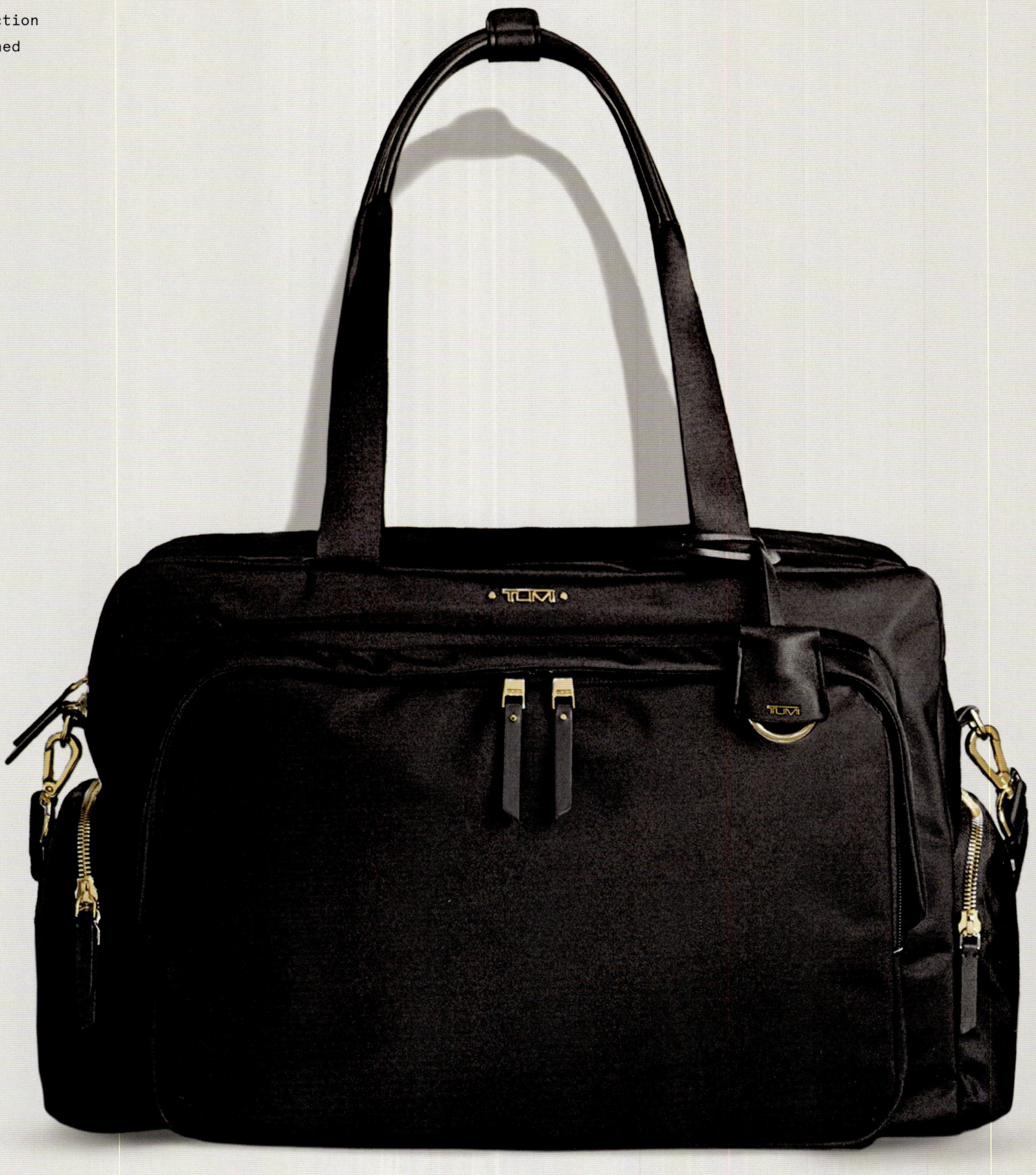

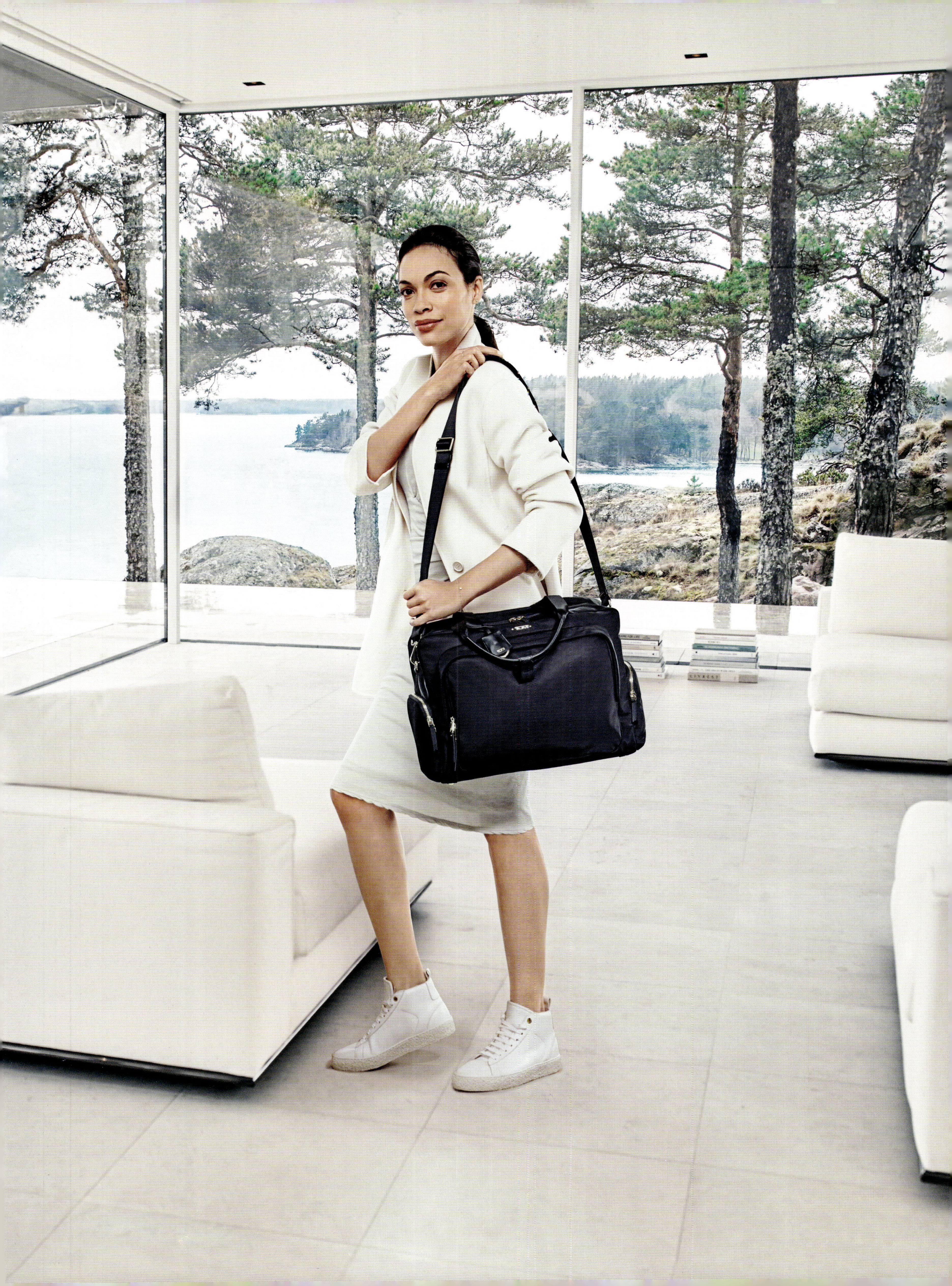

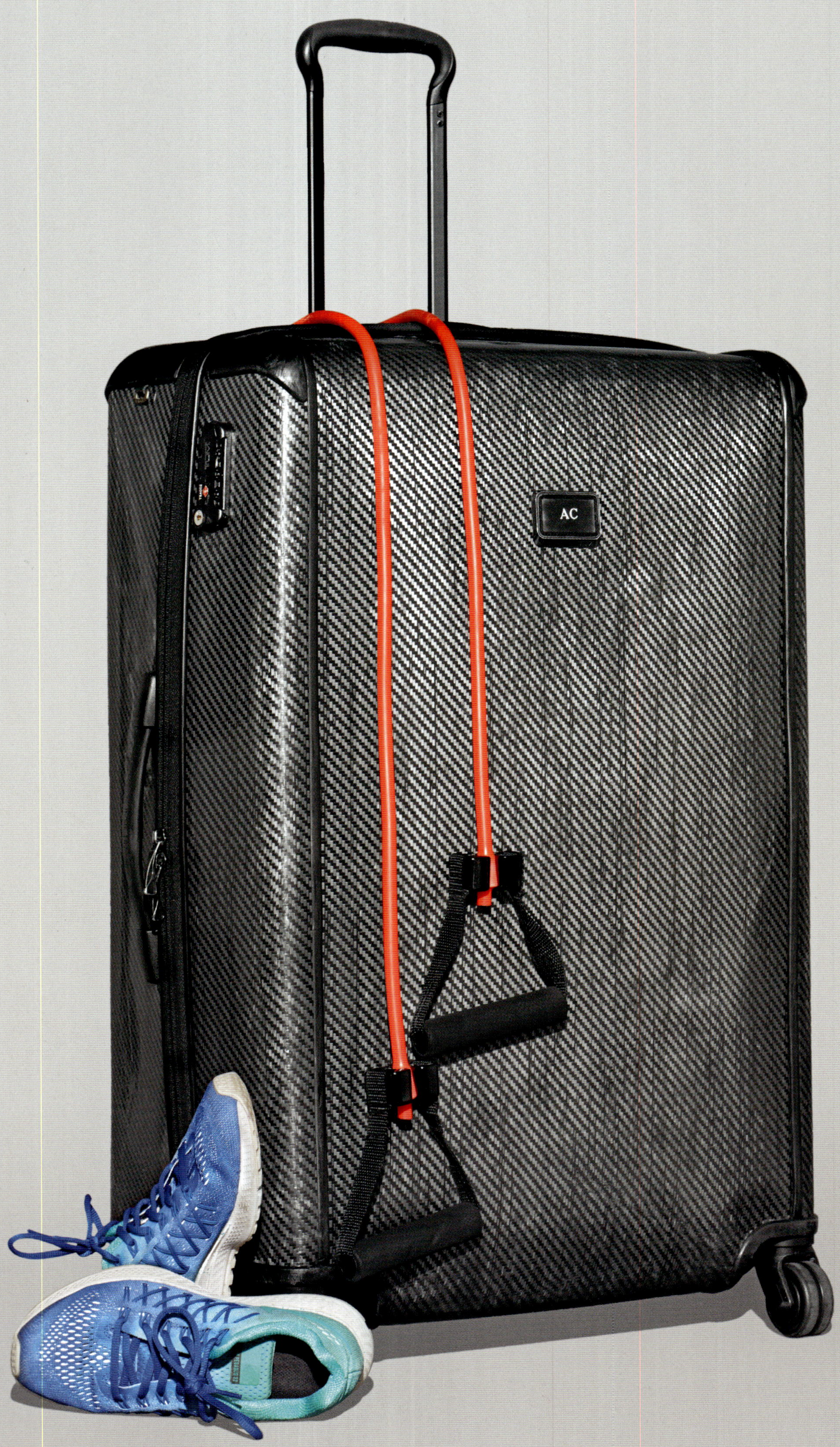
AC

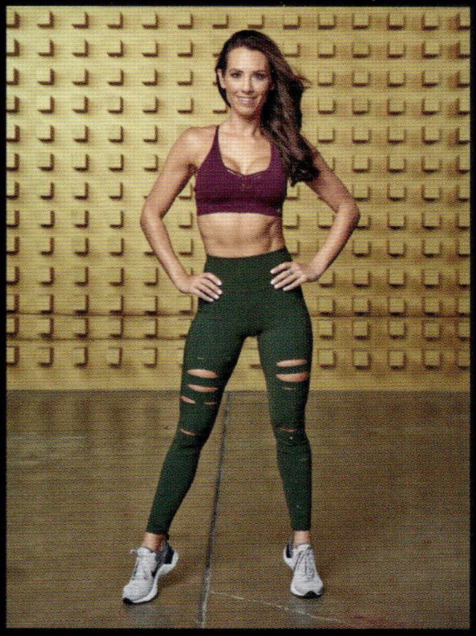

Collection

Tegra-Lite

Style

Extended Ttrip
Packing Case

Collection
Launched

2015

Autumn Calabrese

Fitness + Nutrition Expert and Entrepreneur

As we're running everyone is laughing at me and joking about how annoying my suitcase was.

I remember one particular trip, after I started with Beachbody. Things were going really well and I was on a business trip in Canada with some people from the corporate office. We were late for our flight, so as soon as we got to the airport we just sprinted. We were literally running through the airport, and the luggage I had at that time was old and junky. The wheels were like nails on a chalkboard. And, keep in mind, this wasn't a short run. It took five minutes to get to customs. Then another five to make it to the gate. I was mortified. We wound up missing our flight—we watched them close the door as we approached the gate. When we returned home the next day, I went straight to the mall and bought myself a TUMI suitcase. The biggest one I could find.

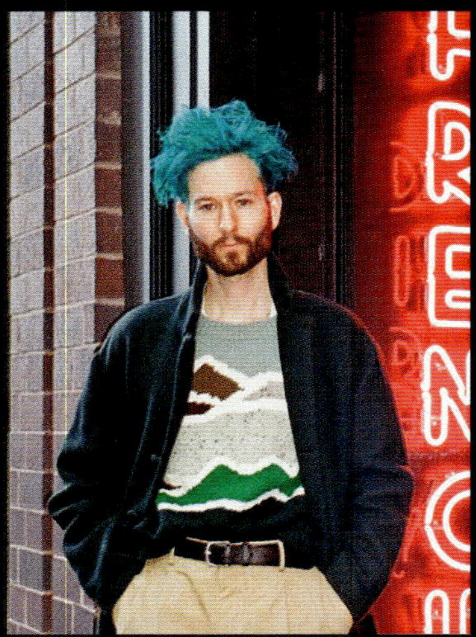

Isaac Hindin-Miller

DJ

Collection
CFX

Style
International Carry-On

Collection
2014

This bag has completely changed my life.

I was one of those people who used to travel with an over-the-shoulder bag. Of course they look good, but they're so unwieldy and inconvenient and they make your shoulders and arms incredibly sore. I also live at the top of a fifth-floor walkup in New York City—there's nothing worse than arriving back at home after a 22-hour flight from New Zealand and carrying your suitcase up five flights of stairs.

When I was younger, I was a skateboarder and I remember going into skate shops and picking up a bunch of stickers. I couldn't always afford a new deck or shoes, but I could always afford stickers. I still pick them up, only now I put them all over my suitcase. I love that this is a serious piece of equipment that's covered in cheap stickers. I love that high/low aspect of it.

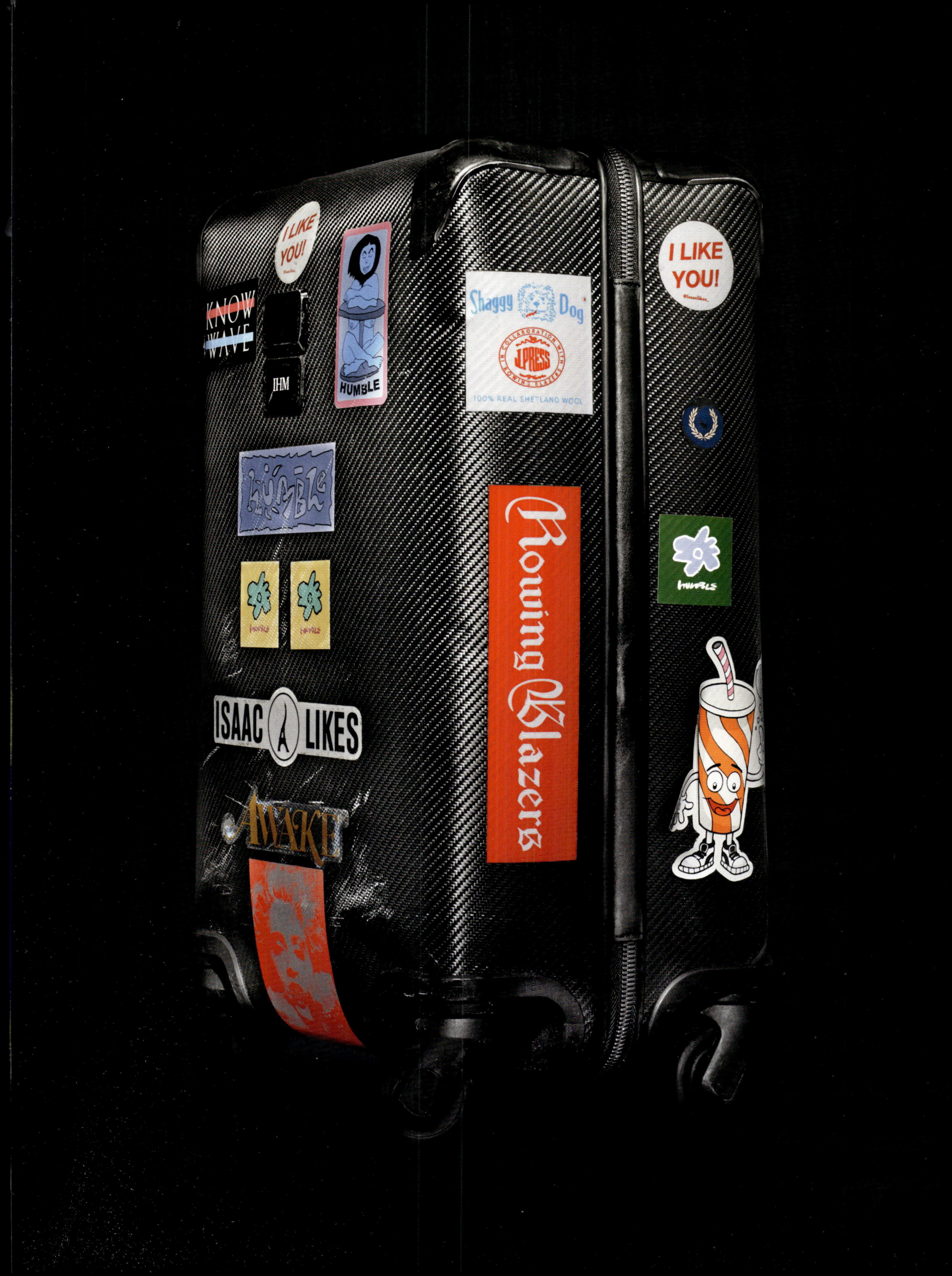
I LIKE YOU!
KNOW WAVE
JHM
HUMBLE
Shaggy Dog
COLLABORATION WITH ROWING BLAZERS
J.PRESS
100% REAL SHETLAND WOOL
I LIKE YOU!
Rowing Blazers
ISAAC LIKES
AWAKE

Collection

Alpha 2

Style

Wheeled Split Duffel

Collection Launched

2014

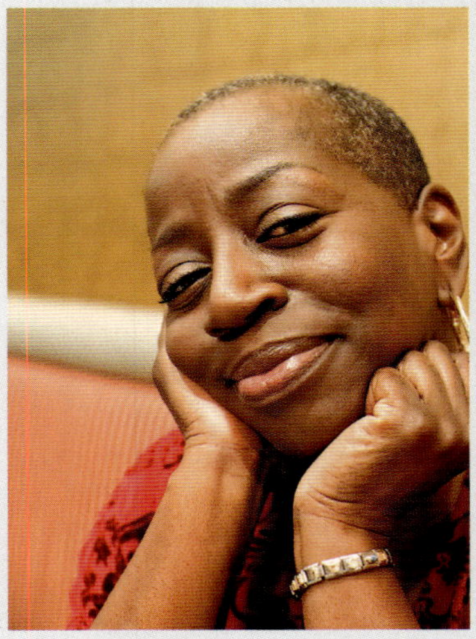

Saundra Williams

Singer and Songwriter

Starr and I would go window shopping all the time while we were on tour with Sharon Jones and The Dap-Kings. It was a great way to chill out and spend time together. While we were out and about, we'd always wind up looking at luggage. We both could not tell you why luggage fascinated us so, but we'd go into travel stores and "ooh" and "ahh" at the luggage. One brand we would "ooh" and "ahh" at all the time was TUMI. We'd be happy to see a TUMI store, no matter what city or country we were in. If they had a TUMI store, we were there.

We knew that the luggage and accessories were high quality, and we would take our time checking out everything, always engaging the salesperson. Well, as fate would have it, Starr and I were at a mall and we got a call that we were about to be gifted with a TUMI bag of our choosing. Needless to say, we were beside ourselves. Actually, we were outside the shopping center waiting for a cab back to our hotel when we got the call. You know what? We canceled that cab and went right back into the mall and into the TUMI store.

Once inside, we saw the collection and we made our choices. We both picked the black Large Split Wheeled Duffel. Starr chose pink accessories (she's the girly-girl) and I chose orange (I'm a Halloween baby).

The day came and I got my bag. I was in love! It was beautiful. Strong, sturdy and easy to push and pull. I immediately started using it as I toured the world with SJDK for the next several years. Man, that bag has been everywhere and it's held up. I had to put some black electrical tape on what I call a "stress area"—the parts of the bag that rub against stairs and escalator steps as I travel. I'm sure with normal use my bag would have been fine but this bag was not being used "normally." I was traveling 10 months out of the year, for years—up and down, trains, planes, vans, and even boats, and she still looks beautiful—black tape and all.

I love my TUMI duffel. She's a tough baby who's been with me through a very special time in my life and career—world tours with Sharon Jones and The Dap-Kings. Our tours are over now and when I look at my bag, I smile as I remember. I use my bag now to store all of my old tour T-shirts, my Daptone Holiday sweater, tour memorabilia, and such, but I rest assured knowing that if I ever had to pull that baby back out to hit the road, she'd be ready to roll.

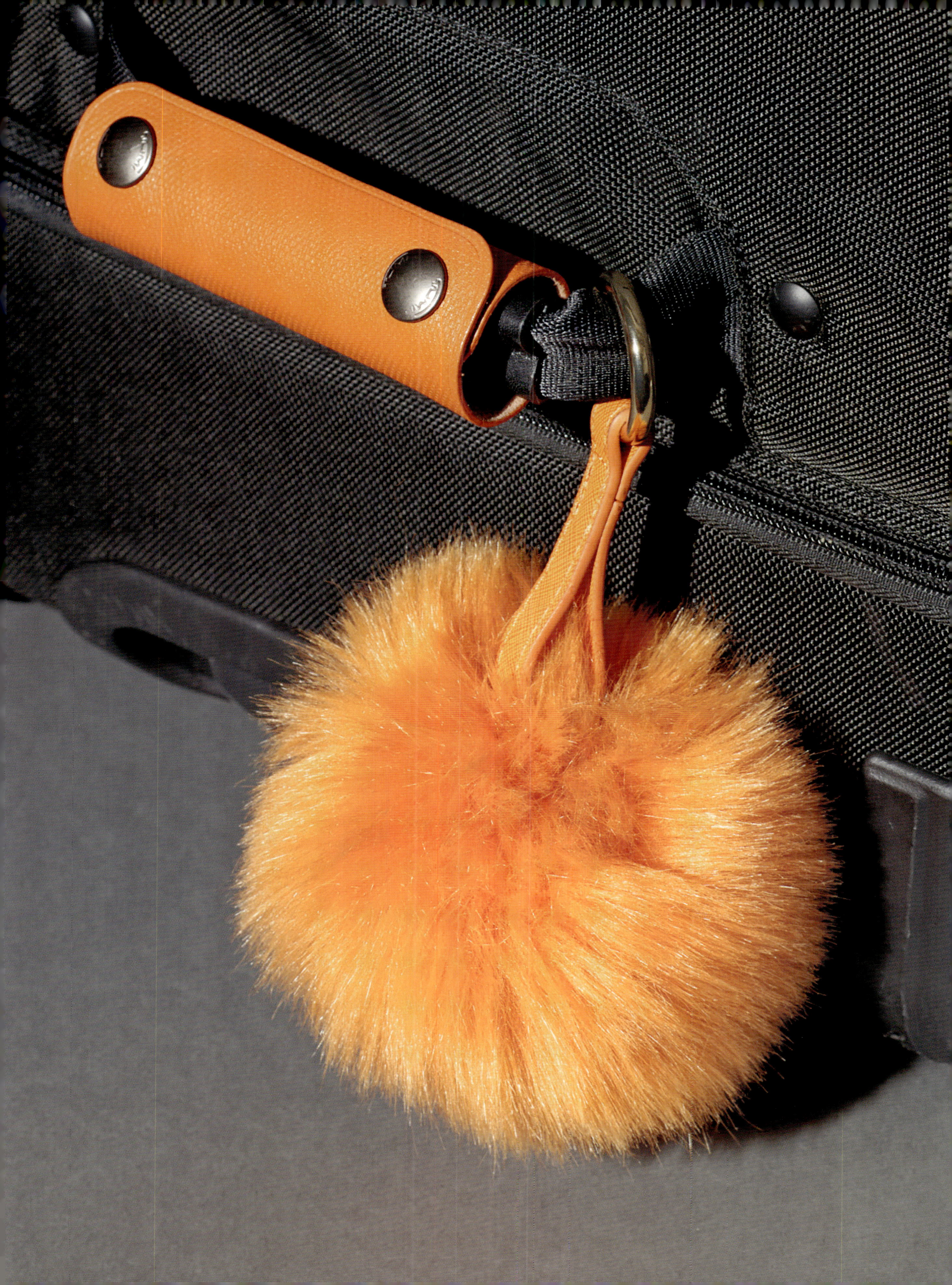

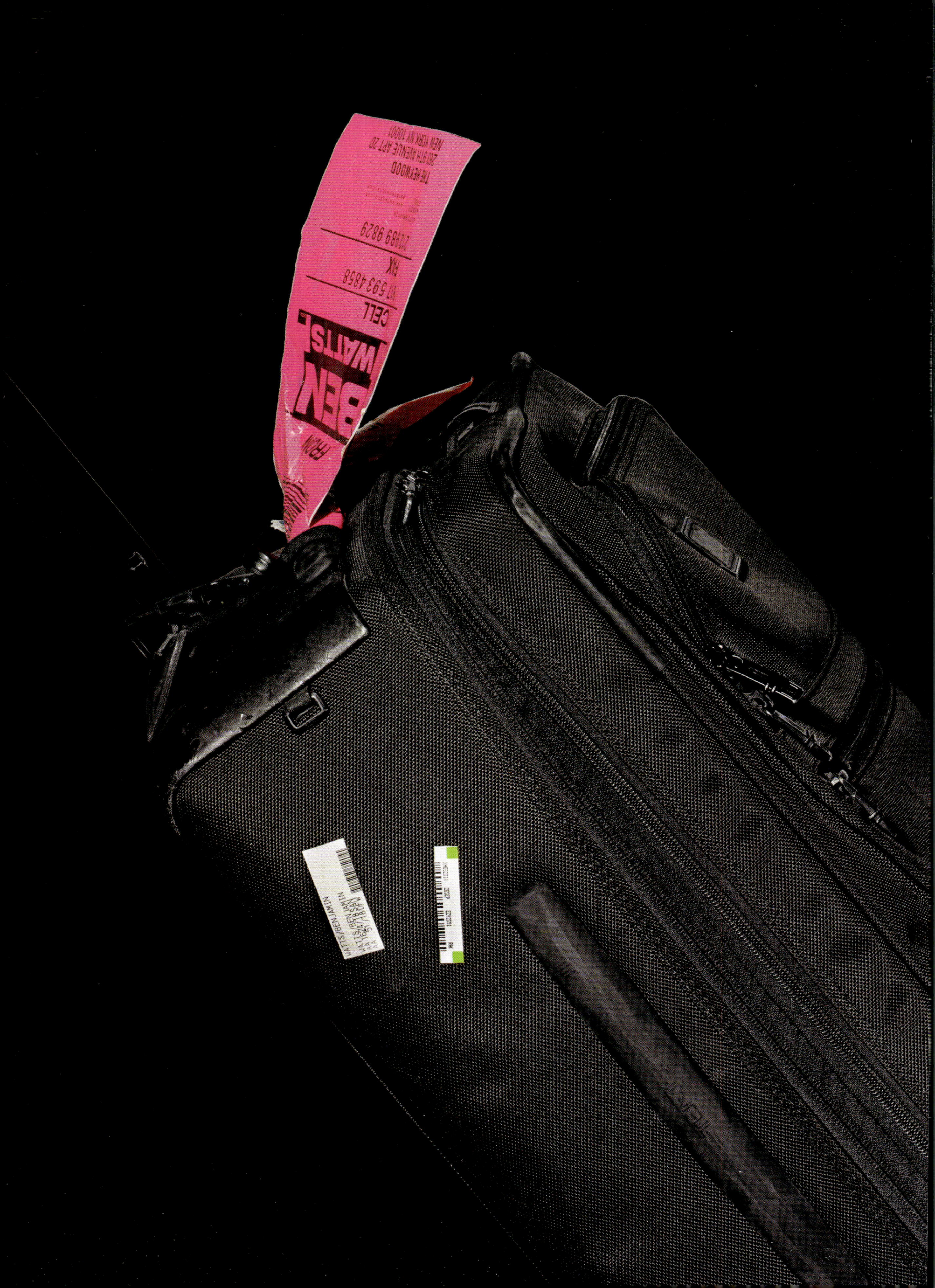
FROM
BEN WATTS.
CELL
FAX
THE HEYWOOD
WATTS/BENJAMIN

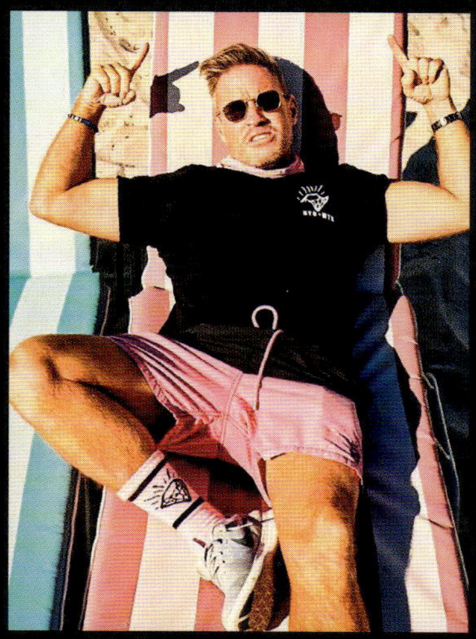

Ben Watts

Photographer

I used to think it was all about what was in the bag, not the bag itself or how well it carried my things.

I was introduced to the brand by my ex-wife in the mid-nineties. I immediately liked it, especially that section on the bottom that held my shirts and shoes. That was ingenious. But then she told me how much it cost and I couldn't believe it. I was young then, and I had the same reaction when we renovated our home. I didn't understand that you need to spend money on good appliances. I thought, "Well, if it's a dryer, it dries." Over time, I came to realize the ease in which my luggage got me from Point A to Point B. Now I notice so many TUMI bags in the overhead compartments of planes, especially in business class. That's because it's made for the educated traveler.

My TUMI challenges me to organize my life and edit things down. It allows me to take only what I need. In many ways, it's a metaphor for life: you have to make choices. When my wife and I split up, we each took our own TUMI luggage. It wasn't even something we discussed. She took hers, and I took mine.

Collection
Alpha 2

Style
Front Lid Short Trip Packing Case

Collection Launched
2014

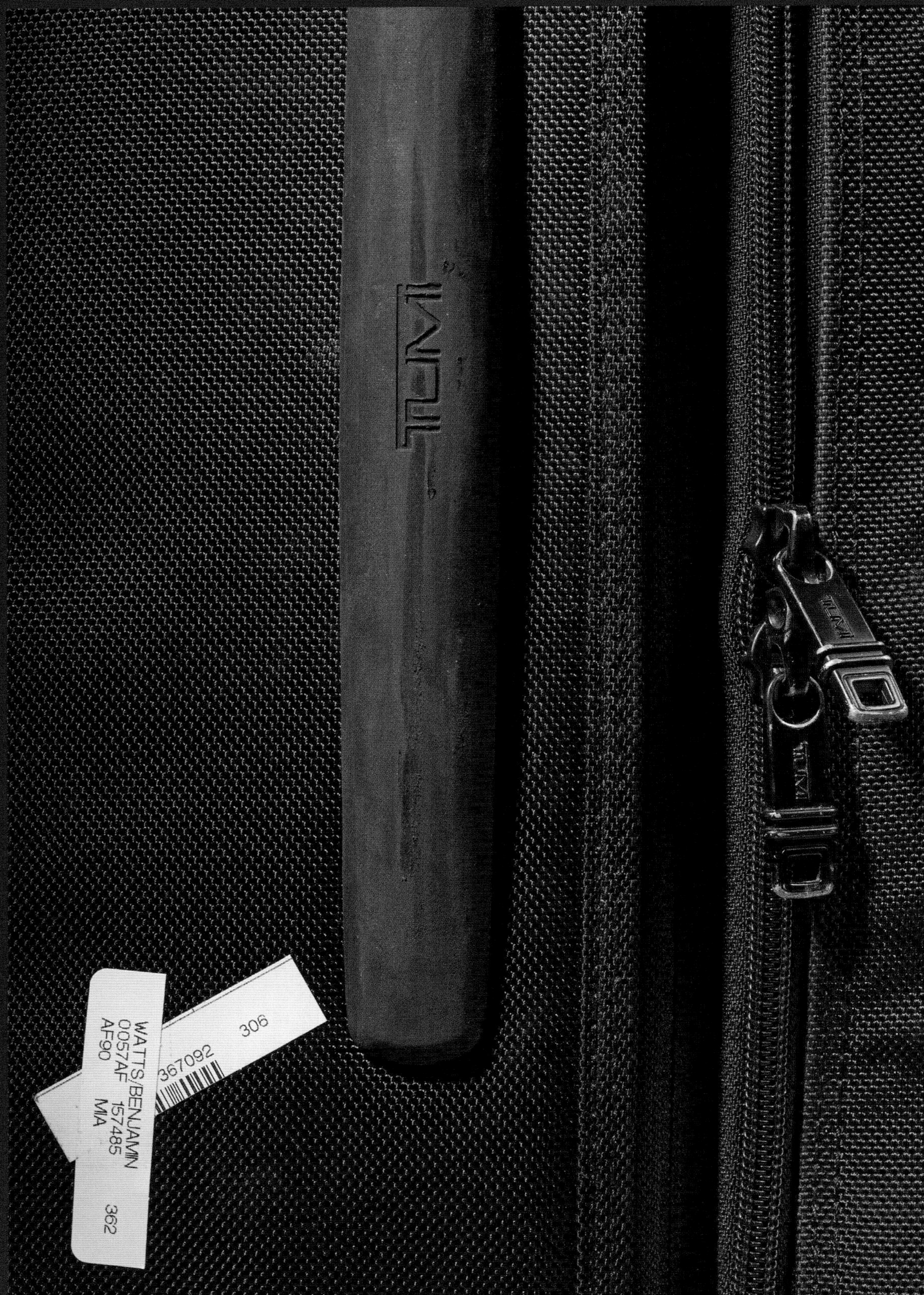
WATTS/BENJAMIN
0057AF 157485
AF90 MIA
362
367092
306

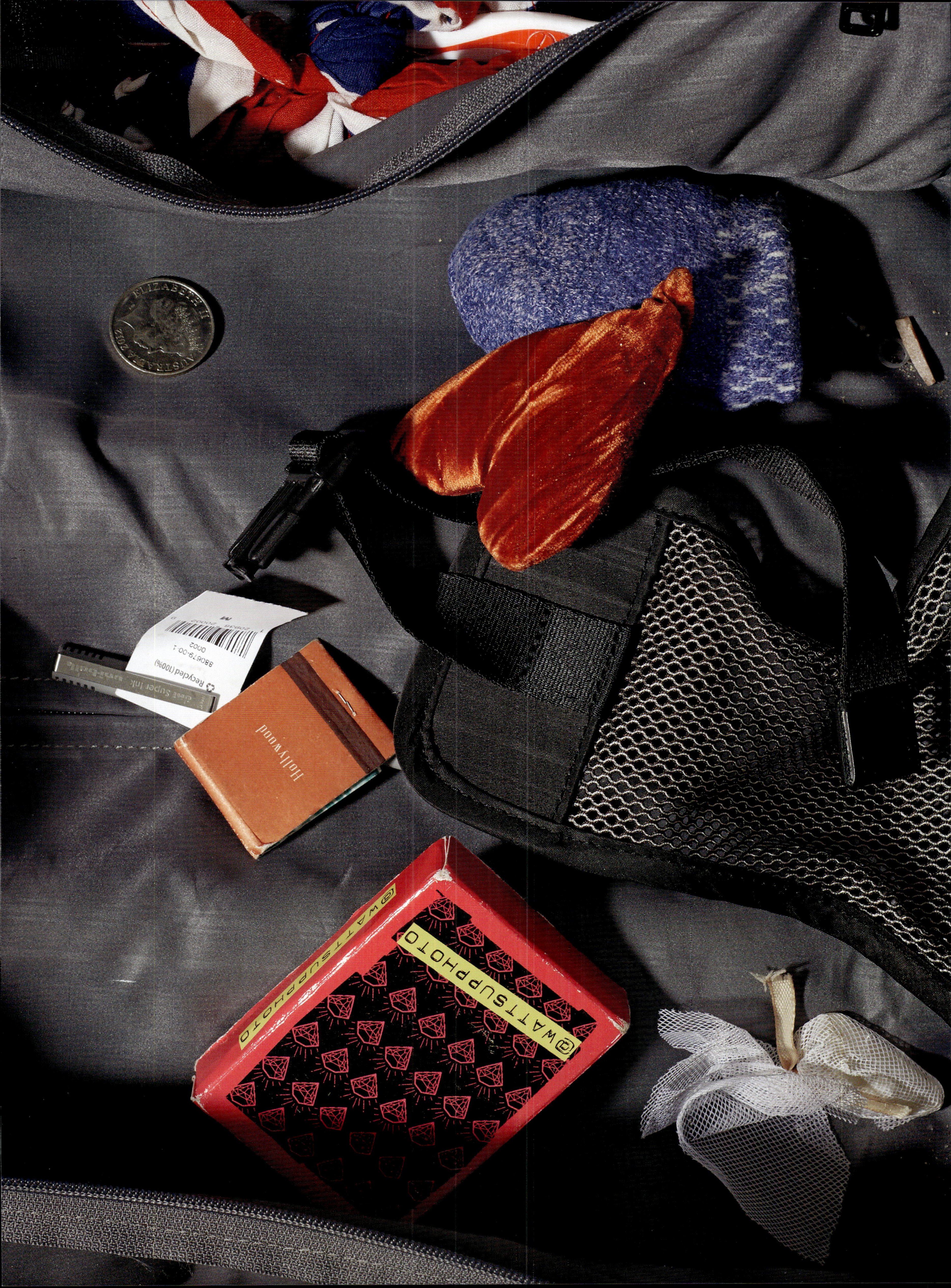
ELIZABETH II
Hollywood
@WATTSUPPHOTO
@WATTSUPPHOTO

You have to realize when you're touring, your luggage gets a lot of wear. You're on planes, trains, and buses every day.

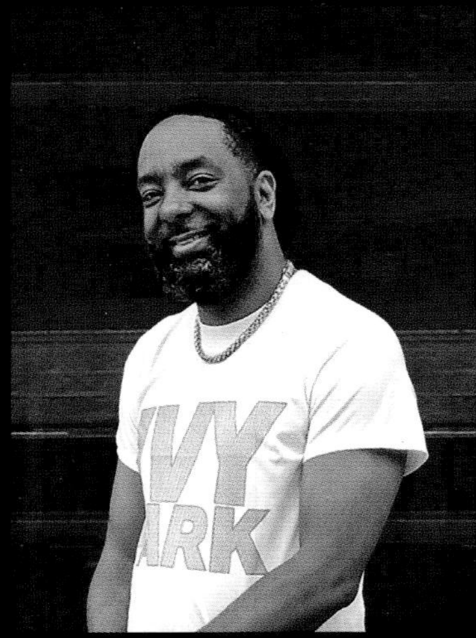

Alan Floyd

Head of Touring Operations for Beyoncé

My dad, Eddie Floyd, is a soul musician. He had a bunch of famous songs, like "Knock on Wood." He used to travel all the time, and he carried TUMI. When I started tour managing, I would literally buy new suitcases every few months, which was becoming expensive. I knew that TUMI had a lifetime guarantee, so I asked my dad, "Is this for real?" And he said it was—even if your bag was jacked up. So I went and bought a suit bag and a trunk, the one with the shoe compartment at the bottom. I thought that was so cool. At the time I was working with Destiny's Child. I remember one time American Airlines ripped the handle off my bag, and I sent it in to TUMI to be repaired. They didn't replace the handle, they sent me a whole new bag. That trunk is a lifesaver. I gave TUMI backpacks to everyone on Beyoncé's crew—dancers, the band—and a lot of them went out and bought luggage to match. A lot of Beyoncé's wardrobe is packed in those TUMI trunks, too. If you're on tour and you don't have this luggage, what are you doing?

Collection
Gen 4

Style
Large
Wheeled
Duffel

Collection
Launched
2005

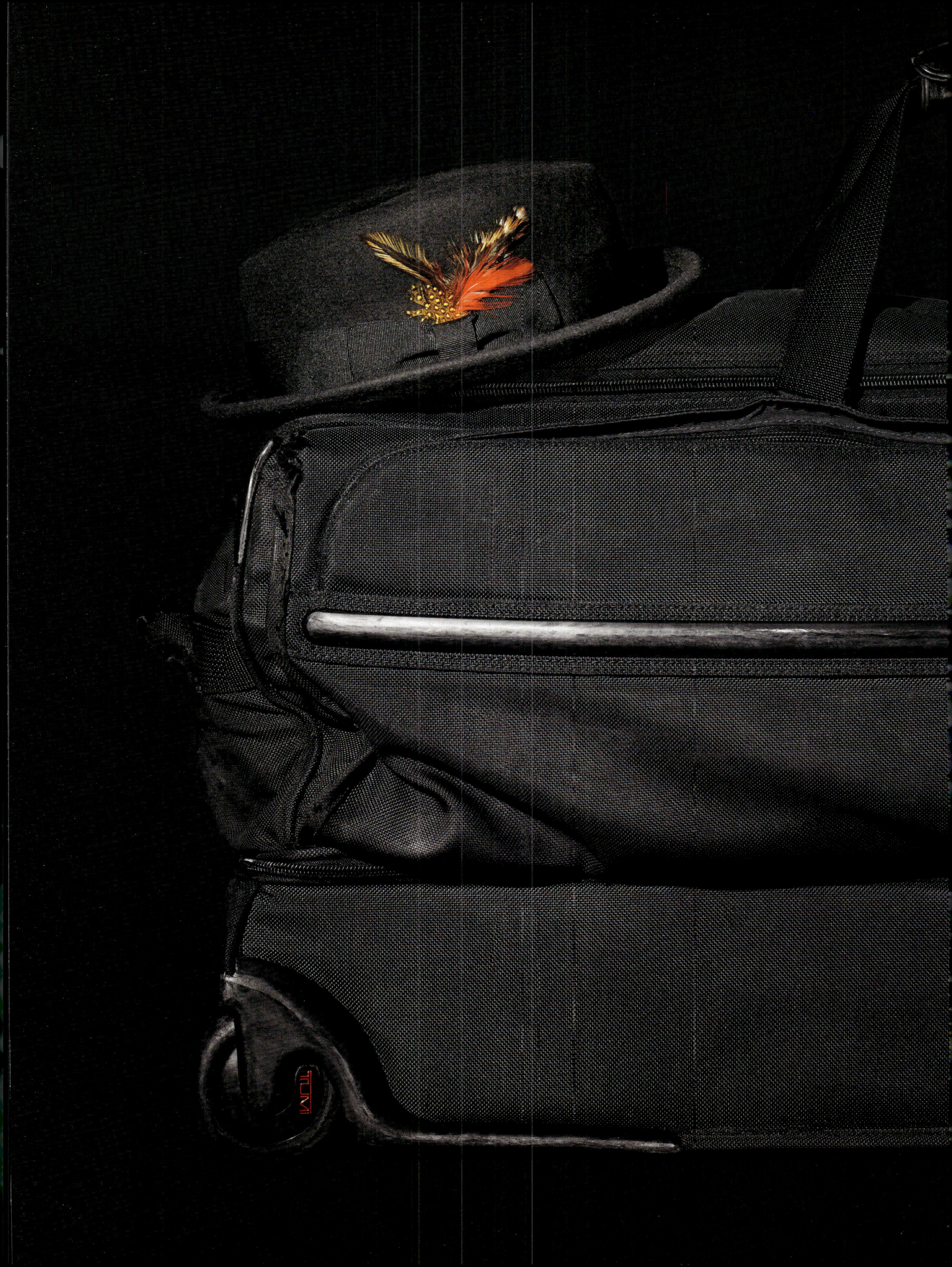
TUMI

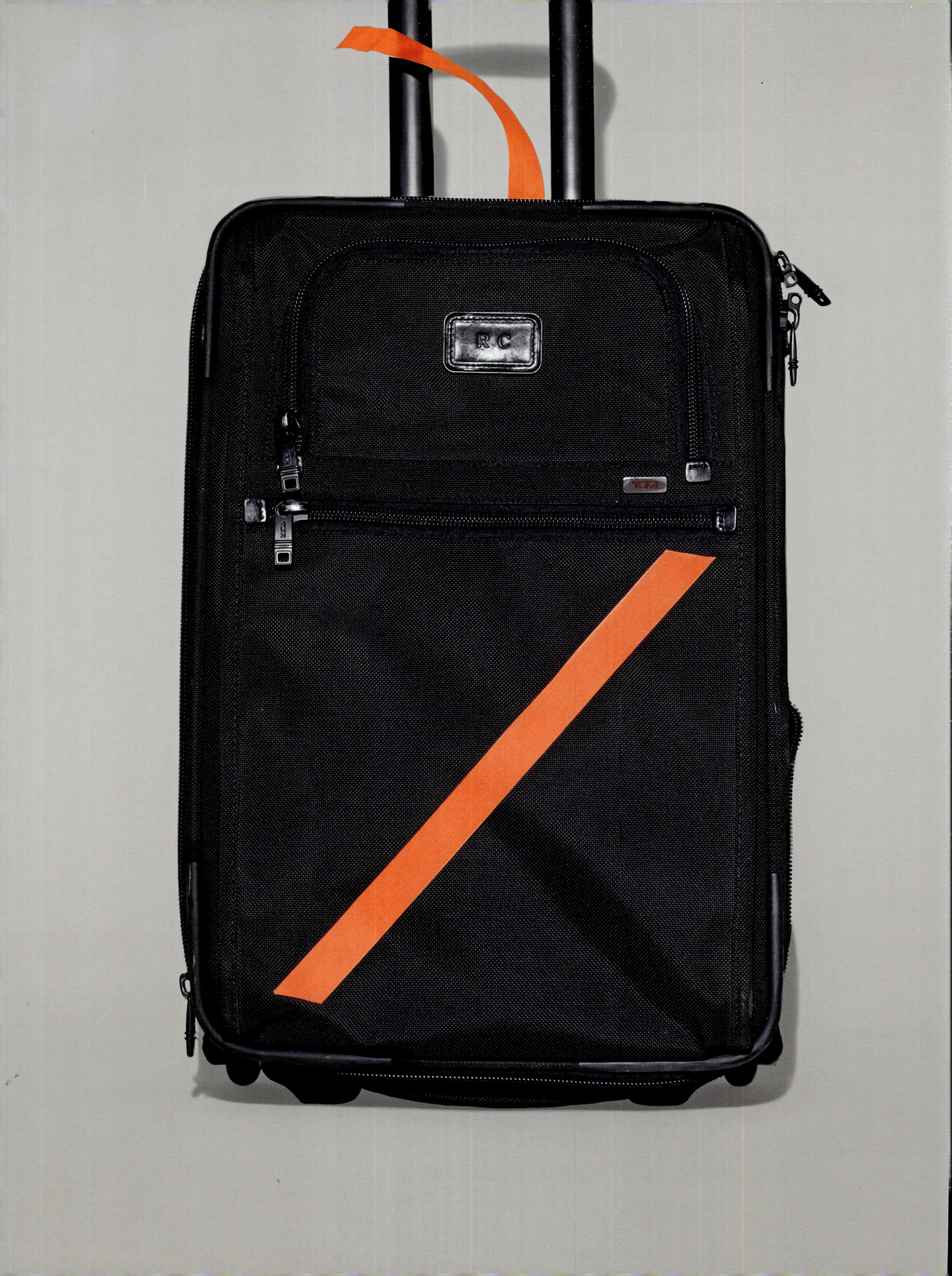
R C

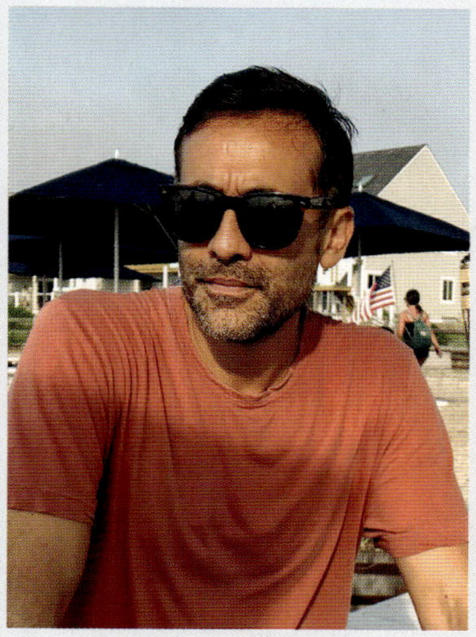

Rodrigo Corral

Graphic Artist

My TUMI is a reminder and reflection of what I value in great objects.

Like restraint. Restraint, to me, says self confidence. That's why I appreciate the simplicity of the logo. It's so reflective of the quality. My first trip with my TUMI was to Mexico about ten years ago.

I left the cab with my carry-on, but forgot my TUMI in the trunk. If you've ever been to Tulum, you know there's one main road in and out of the town. I knew my cabbie would be driving past me again, so I spent the next two hours flagging down every cab heading out of Tulum. I stopped about 20 of them—thank God I speak Spanish. When I found my guy, he was pretty puzzled—he had no idea my bag was in his car. He said, "You're the guy who's been causing all this traffic?" But you know, an object of that quality, it almost has a presence to it. My mother always had a beautiful set of luggage, and there was a sense of pride in how she presented herself when she traveled. And I have that same feeling. My bag is now my family's bag. One day it will belong to one of my three children. Something that's been passed down possesses stories. You get to have your own dialogue with the object, and then the story expands and evolves over time. For now, though, my bag will live through many more stories with me.

Collection

Alpha Legacy

Style

International Carry-On

Collection Launched

2008

I love things that have a story. Now my bag tells my life story.

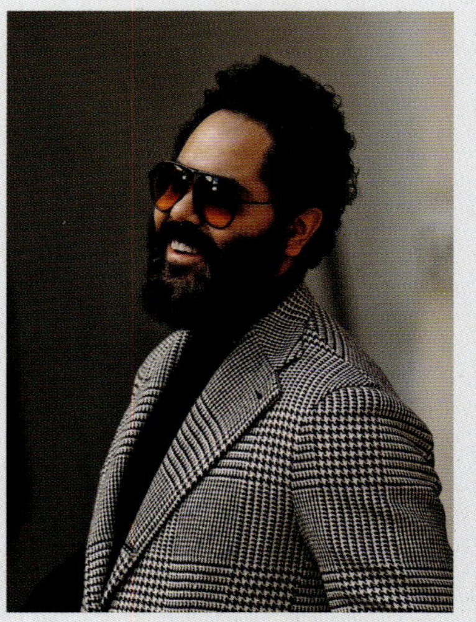

Angel Ramos

Founder and Chief Creative Officer of 18th Amendment

I went to The High School of Art and Design. So many legendary artists and designers went there: Marc Jacobs, Tony Bennett, Calvin Klein—and some of the most iconic graffiti writers in New York City, like Lady Pink. I've always been intrigued by imperfection. It's the arc for everything I do. So a bunch of buddies and I graffitied over my TUMI bag. Just because something has been painted on doesn't mean it's been vandalized or defaced. This is how an artist tells his story. This is a city boy's monogram. It says, 'he's from Brooklyn. He's an artist.'"

Collection
T-Tech Cargo

Style
Medium Trip Packing Case

Collection Launched
2013

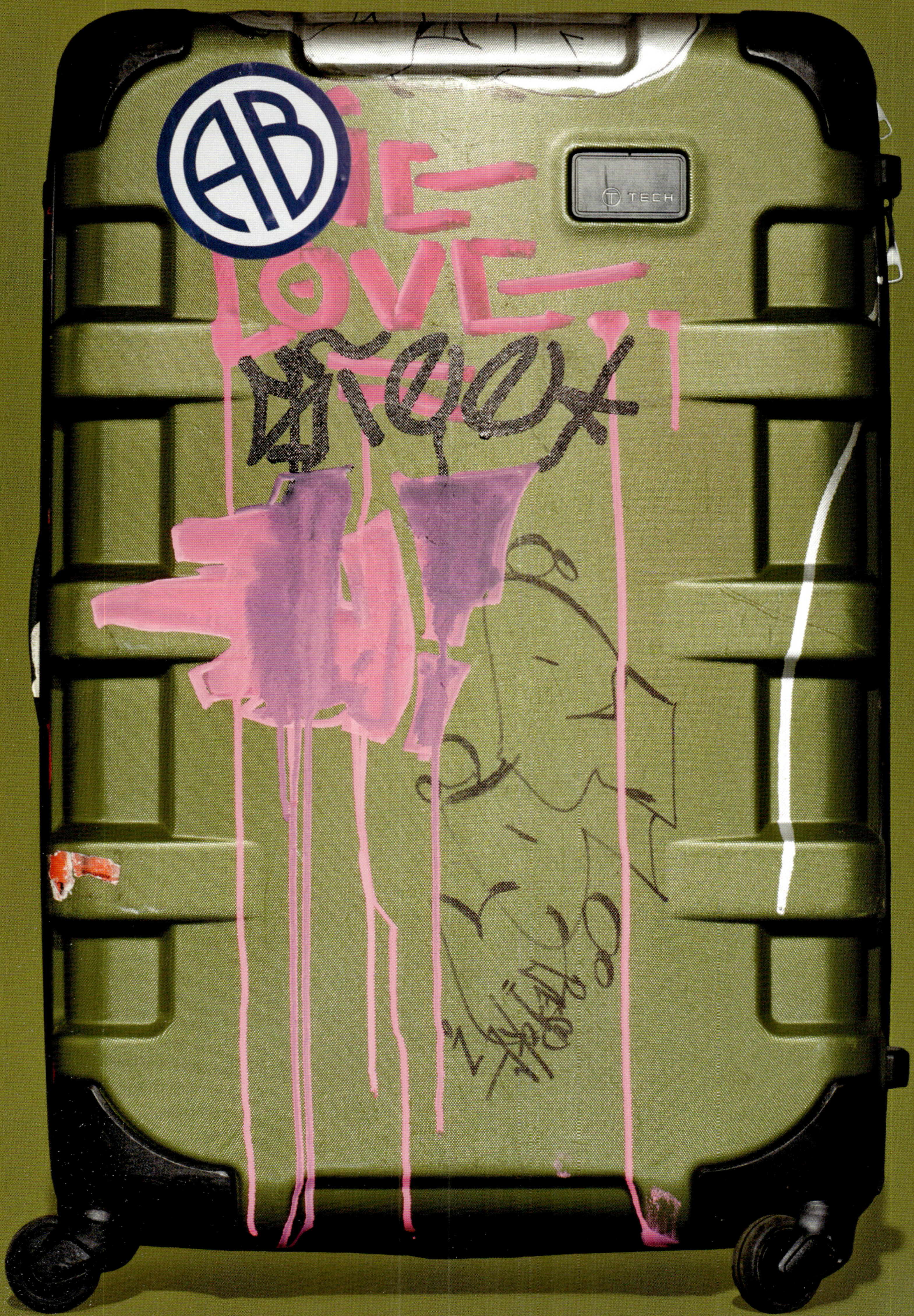
AB
TECH
LOVE

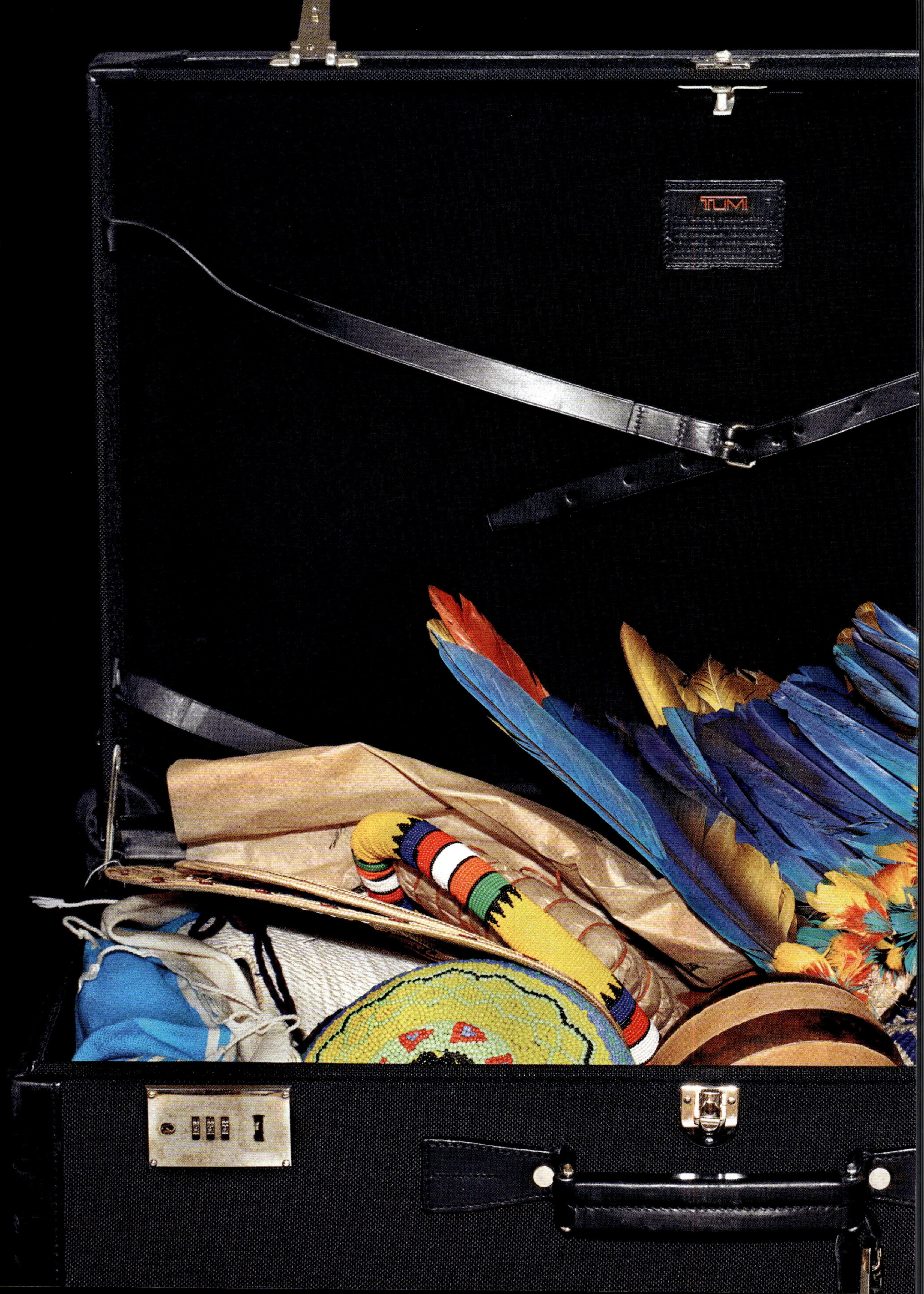
TUMI

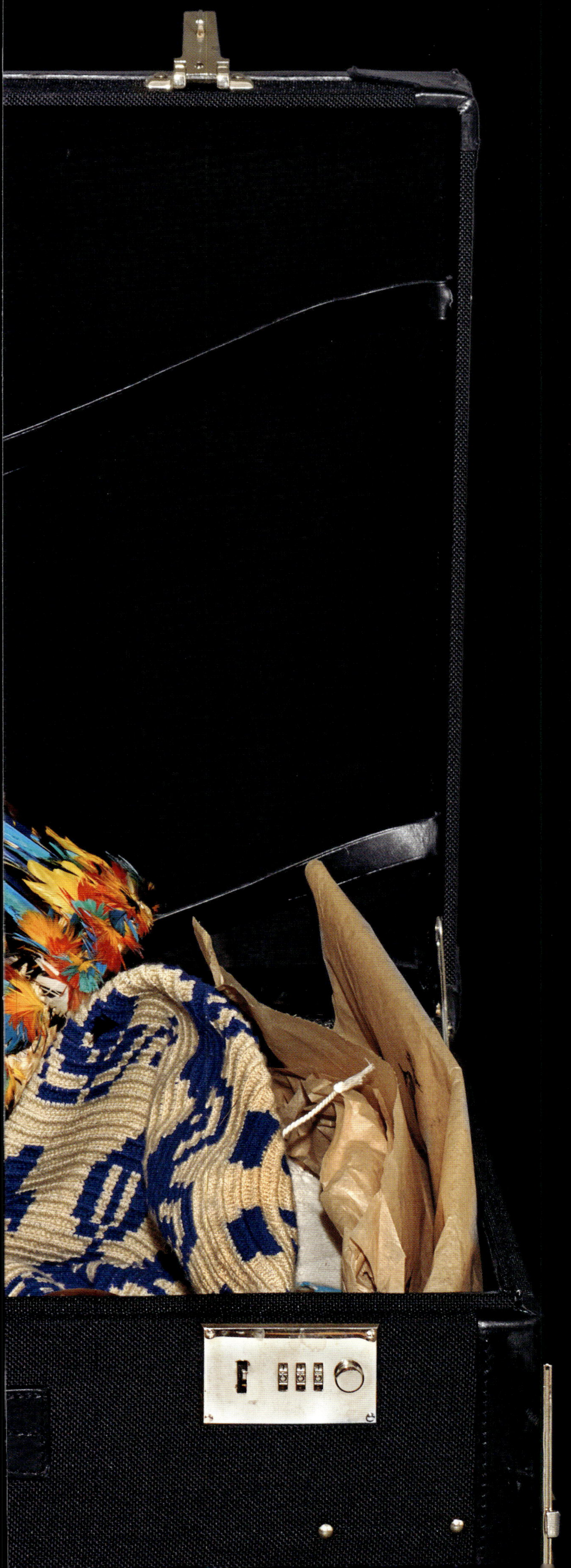

Collection

Townhouse

Style

Stanhope Trunk

Collection Launched

2008

Ruven Afanador

Photographer

As a photographer, I have to travel often, and for the longest time, I was having so many problems with my luggage. Whether it was super expensive or inexpensive, it would barely make the trip. No matter what I tried, something always happened. It became a comical thing. One time I was working on Madison Avenue and I saw a TUMI store. I remember reading that they had a new designer, someone who was reinventing the brand. So I went in. I really knew very little about this brand but then I saw this trunk. It was so beautifully constructed. It had perfect edges. I was so impressed. It was exactly what I wanted.

I used that trunk for more than ten years. Because I'm a photographer, I get to go to pretty exotic places. I collect everything from custom jewelry from Berber people or animal teeth. You know, when I compare it to my cameras, cameras would break and let me down—and they're always changing, there's always a better one. But my trunk never let me down. The trunk was a different thing. I'm very endeared to it. Right now it's in storage filled with those treasures that are very special to me. And I don't plan on ever getting rid of it.

I'm always collecting treasures from my trips. This trunk has carried so many of them for me.

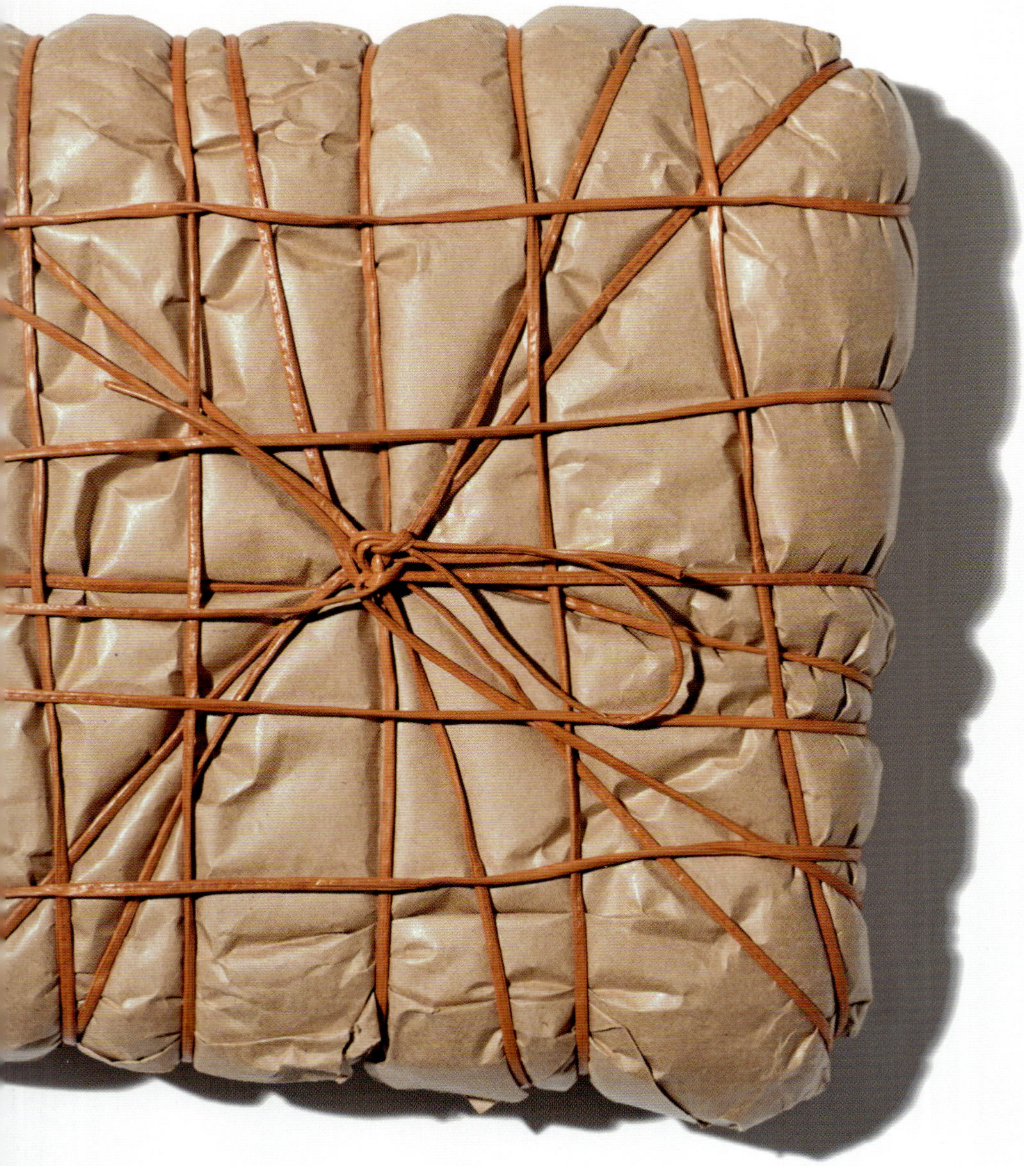

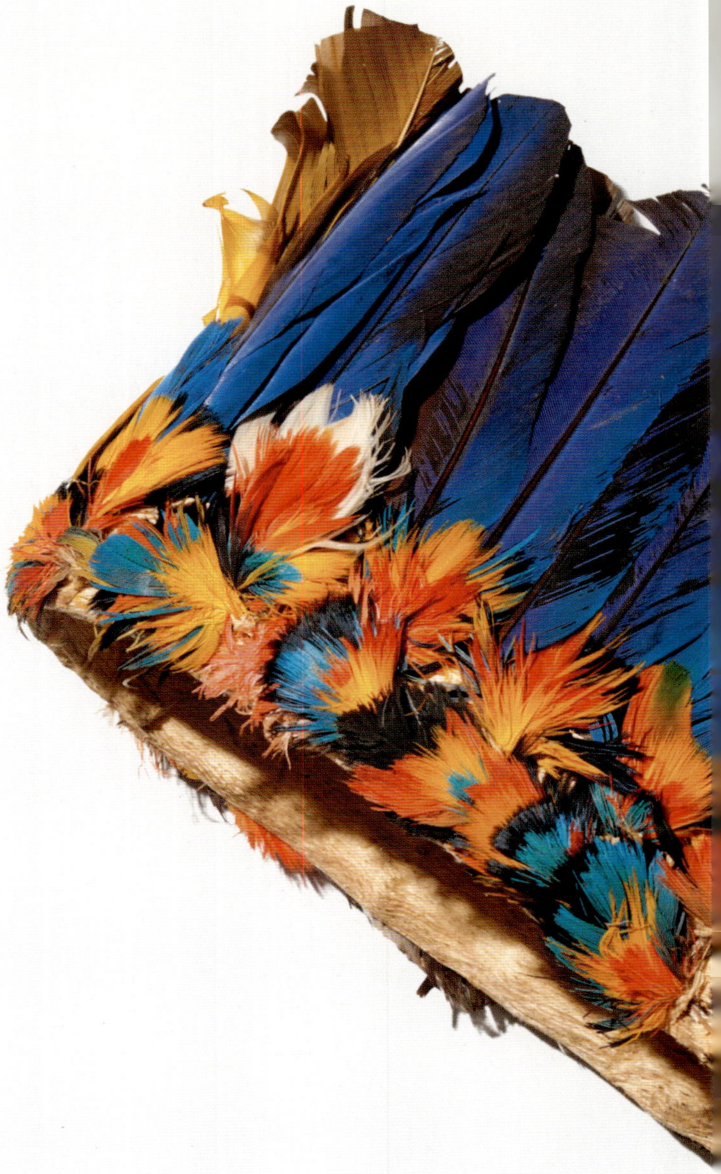

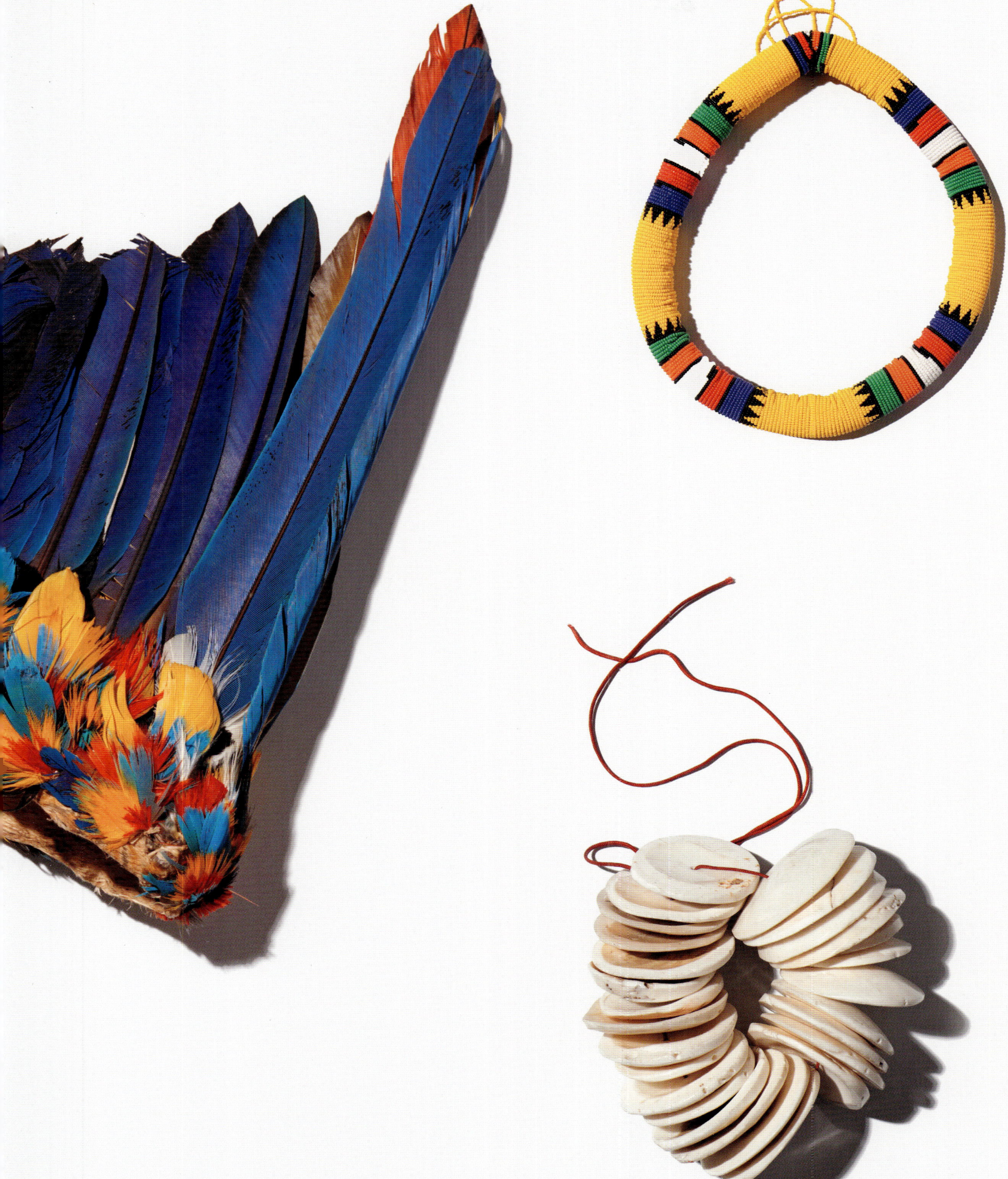

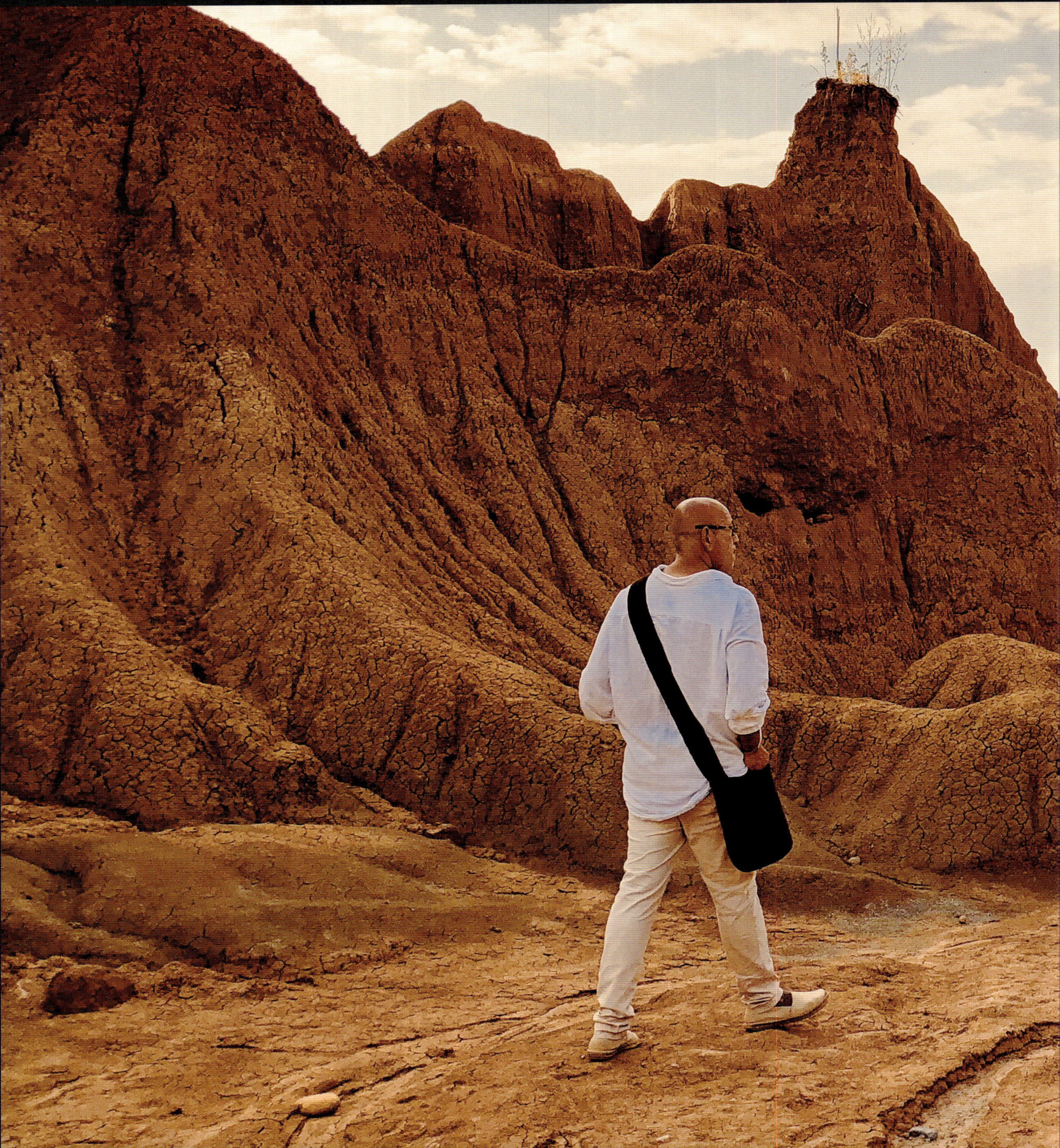

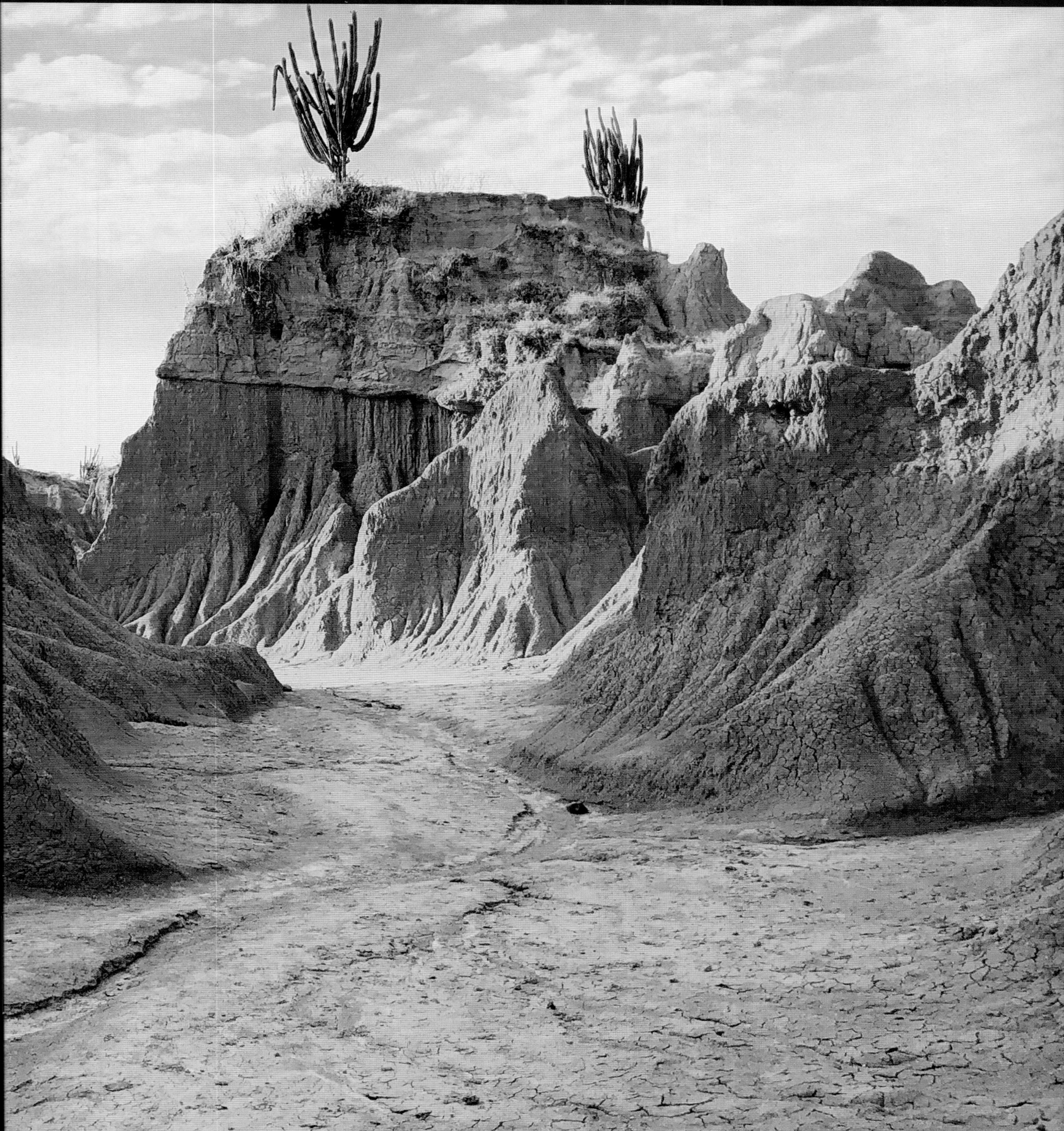

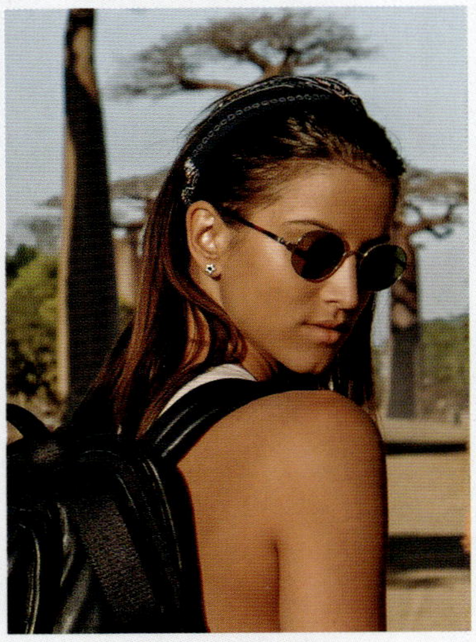

Lexie Alford

World Traveler

Collection
Alpha 2

Style
Brief Pack

Collection Launched
2014

My TUMI has been to 75 countries out of the 196 I've been to.

I spent my 20th birthday traveling solo in Jordan. I went to see Petra by night, and I had only my TUMI backpack with me. I ran out to the middle of the desert and it was pitch black. I was all by myself and two hours late. As I turned the corner in the canyon, I ran into a Berber tribesman. He said, "Calm down, don't worry. You haven't missed anything." He said he lived in some caves behind Petra where nomadic Berbers have lived for hundreds or thousands of years. He noticed I was a photographer—I carried my camera and a tripod in my backpack—so he took me farther into the canyons to take photos. He took me to this amazing viewpoint, then slept at a canyon overlooking Petra.

At some point I was on the back of a donkey with him, wearing my TUMI backpack, and he was serenading me with Jordanian songs. I was being romanced by this desert nomad. It was one of the most beautiful experiences of my life. My TUMI has made me feel more comfortable in all the crazy places I've been. It's all that I have connecting me to my life at home.

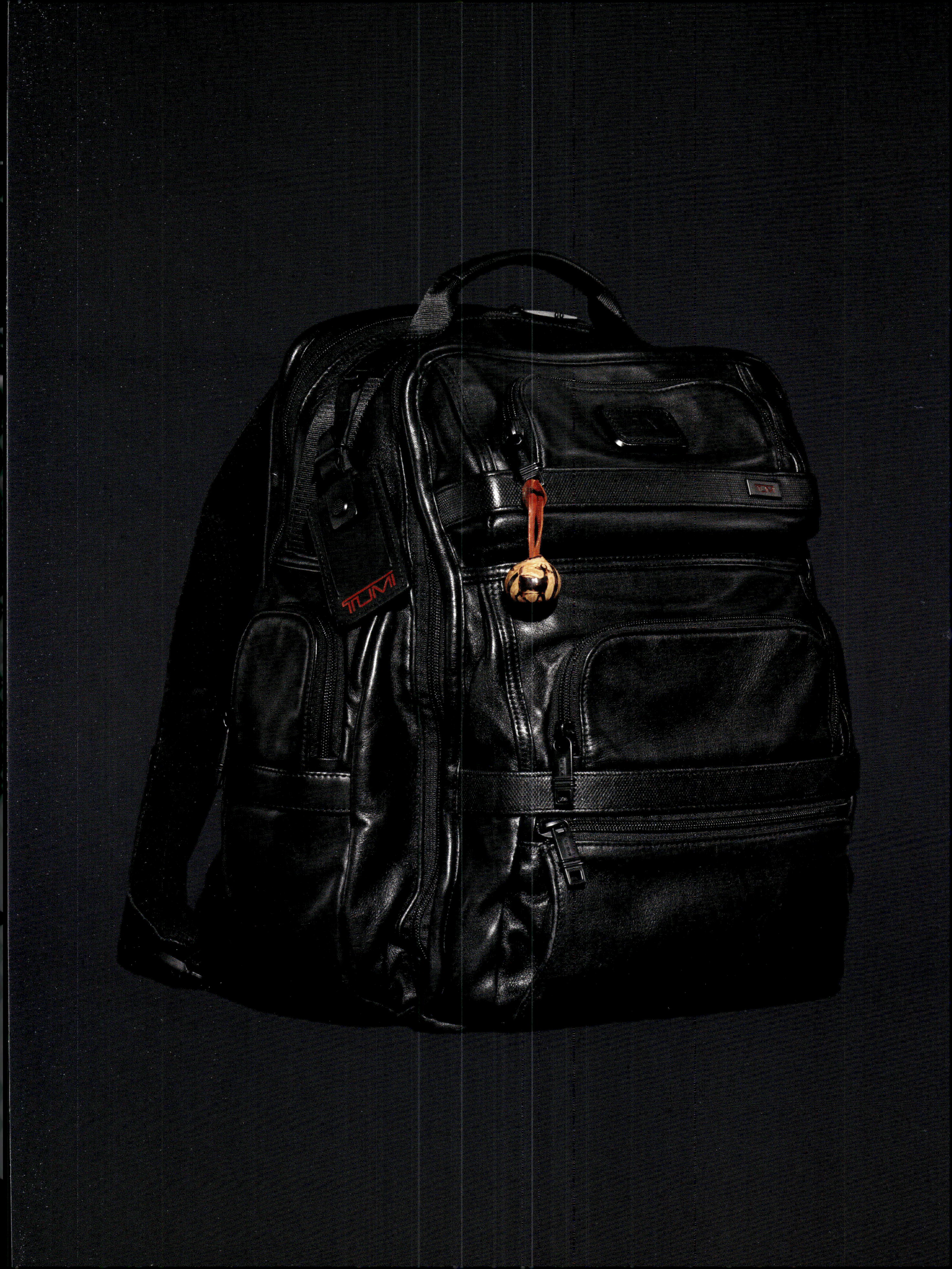
TUMI

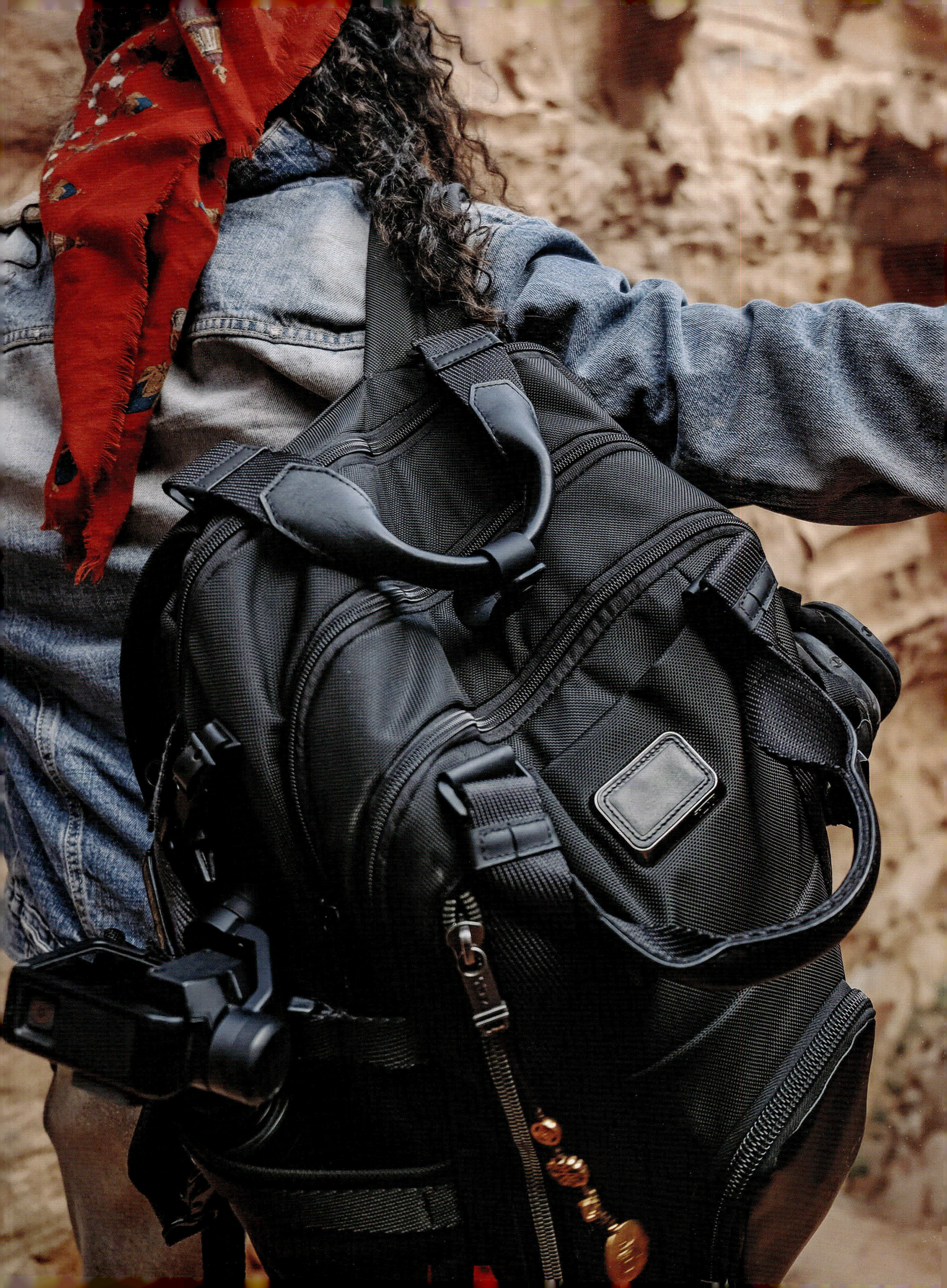

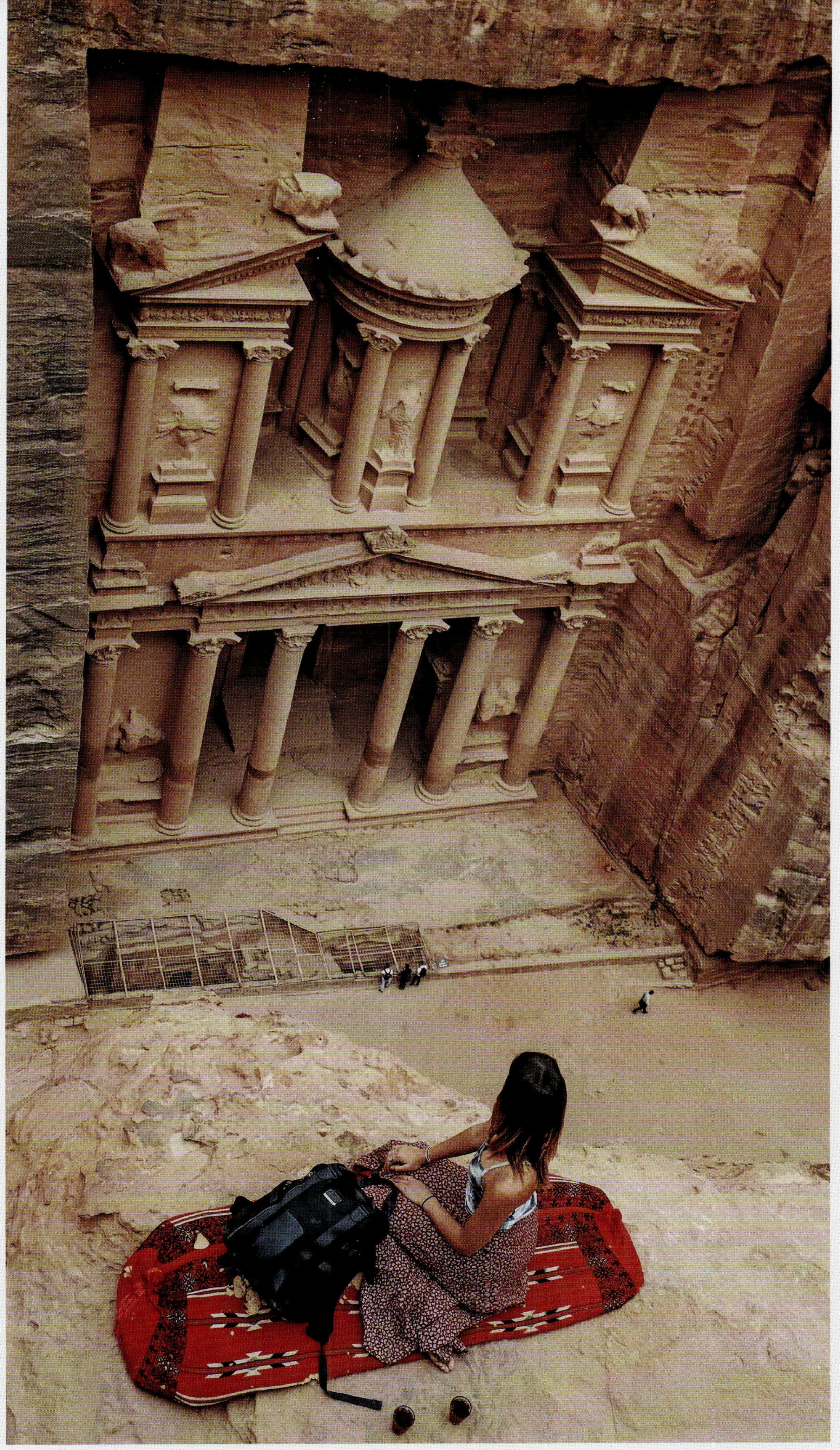

Collection

Alpha 2

Style

Travel Satchel

Collection
Launched

2014

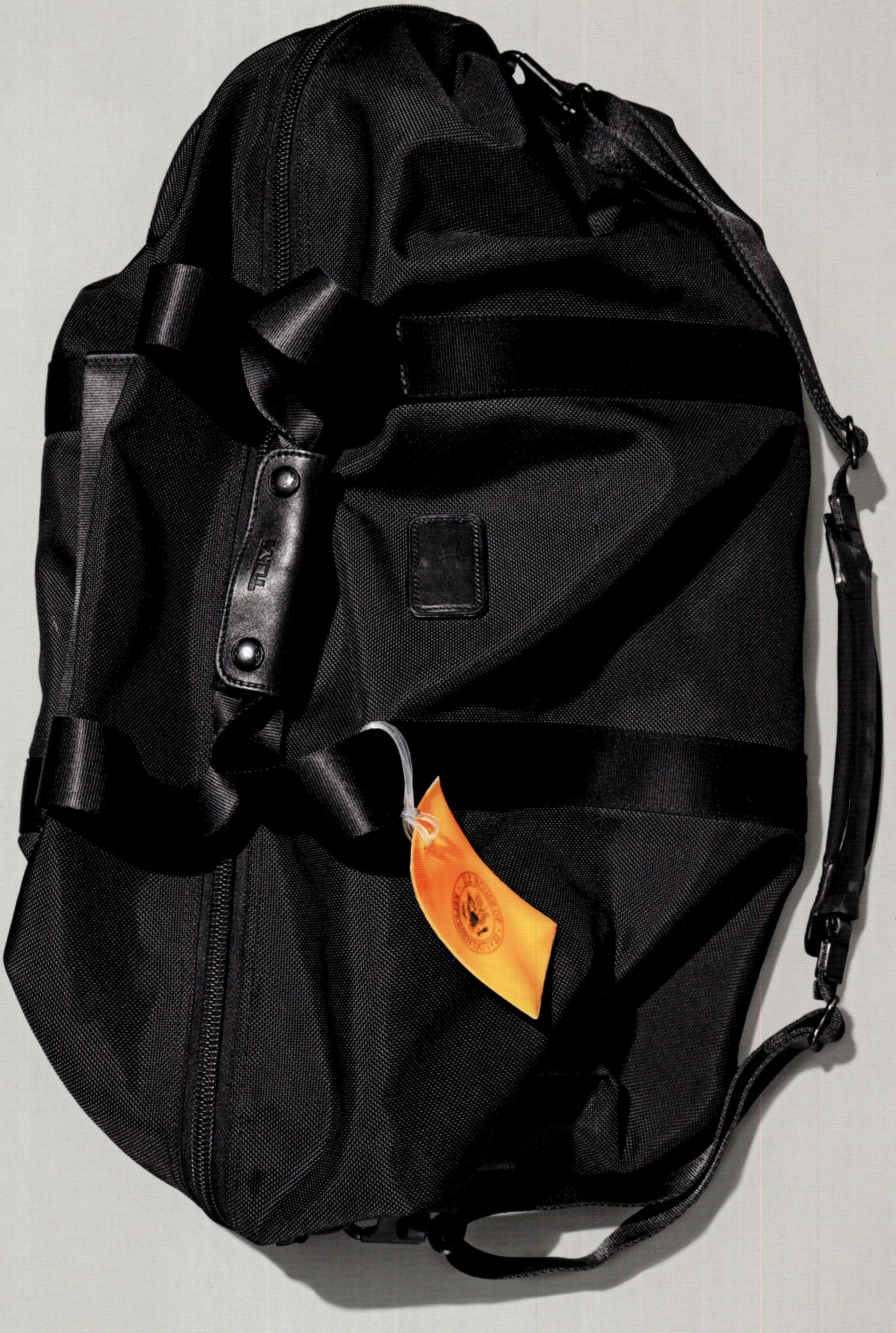

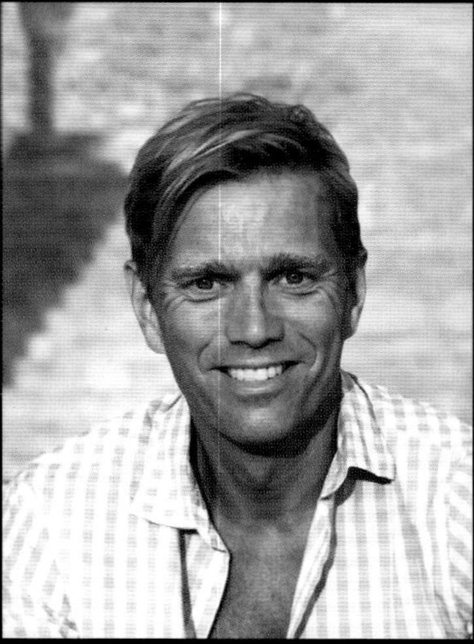

The TUMI Collection

First published in the United States of America in 2021 by
Rizzoli International Publications, Inc.
300 Park Avenue South New York, NY 10010
www.rizzoliusa.com

Editor: Matt Hranek
Photographer: Stephen Lewis
Design: Rodrigo Corral Studio, Inc.
Art Director & Designer: Chris Allen
Art Director: Rodrigo Corral
Art Director: Anna Corral
Retoucher: Derick Arjune
Design Assistant: Dasha Buduchina

Publisher: Charles Miers
Associate Publisher: Anthony Petrillose
Editor: Gisela Aguilar
Design Coordinator: Olivia Russin
Production Manager: Alyn Evans
Managing Editor: Lynn Scrabis

Printed in China

2021 2022 2023 2024 / 10 9 8 7 6 5 4 3 2 1
ISBN: 978-0-8478-6455-3
Library of Congress Control Number: 2020903250

Visit Rizzoli online:
Facebook.com/RizzoliNewYork
Twitter: @Rizzoli_Books
Instagram.com/RizzoliBooks
Pinterest.com/RizzoliBooks
Youtube.com/user/RizzoliNY
Issuu.com/Rizzoli

Photo Credits

Milad Abedi: page 130; Ruven Afanador: page 112-113, 115, 136-137; Courtesy of Lexie Alford: page 138, 140-141; Courtesy of Glenda Bailey: page 33; Zach Beeker: page 30-31; Jamie Cary: page 26, 28, 29; Anna Corral: page 129; Pete Don: page 68; Courtesy of Cristina Ehrlich: page 65; Thomas Falcone: page 42-43; Courtesy of Jennifer Fisher: page 52, 54; Courtesy of Randy Florke: page 143; Courtesy of Alan Floyd: page 126; Courtesy of Melody Gardot: page 94; Anthony Geathers: page 85; Alberto Gonzalez: page 60; Dylan Gordon: page 78-79; Herring & Herring: page 105; Courtesy of Isaac Hindin-Miller: page 118; Max Hoell: page 77; Taylor Jewell: page 111; Courtesy of Claudia Kim: page 96; Darren Lau: page 73; Stephen Lewis: page 16, 21, 24, 32, 34, 35, 44, 46, 49-51, 53, 55, 56, 58, 59, 61, 64, 66, 69-71, 72, 74, 75, 80, 83, 84, 87, 90, 93, 97, 98, 100 (bottom), 101, 103, 104, 106, 107, 109, 110, 116, 119, 121, 122, 124-125, 127, 128, 131, 132-135, 139, 142; Courtesy of Deborah Lippmann: page 92; Jennifer Livingston / Trunk Archive: page 19; Courtesy of Mellian Group, Inc: page 57; Courtesy of Charlotte McKinney: page 100 (top); Courtesy of Kathryn Minshew: page 108; Courtesy of Petra Nemcova: page 17-18; Serge Normant: page 82; Courtesy of Anita Patrickson: page 86; Jenna Peffley: page 102; Per Bernal: page 117; Rambo: page 20, 22, 23; Courtesy of Nikki Reed: page 25; Courtesy of Jon Rose: page 81; Adam Ross: page 88, 89; Courtesy of Maia & Alex Shibutani: page 48; Alan Silfen: page 12, 14, 15, 36, 39, 40, 41; Franco Tettamanti: page 95, 38, 114; Michael Turek: page 45, 47; Courtesy of Ben Watts: page 123; Dan Williams: page 120; Jesse Yonover: page 62-63; Courtesy of Joe Zee: page 91

Visit TUMI online:
Facebook.com/TUMITravel
Twitter: @TUMITravel
Instagram.com/TUMITravel
Pinterest.com/TUMITravel
Youtube.com/ TUMITravel